BUILDING
with
MASONRY

BUILDING
with
MASONRY

Richard Kreh

The Taunton Press

Taunton
BOOKS & VIDEOS
for fellow enthusiasts

Printed in the United States of America
10 9 8 7 6 5 4 3 2

For Pros / By Pros™: Building with Masonry was
originally published in 1998 by The Taunton Press, Inc.

For Pros / By Pros™ is a trademark of The Taunton Press, Inc.,
registered in the U.S. Patent and Trademark Office.

The Taunton Press, Inc., 63 South Main Street, PO Box 5506,
Newtown, CT 06470-5506
e-mail: tp@taunton.com

1/31/03   OCLC   693.1
Kre

Library of Congress Cataloging-in-Publication Data

Kreh, R. T.
    For Pros / By Pros™: Building with Masonry / Richard Kreh.
      p. cm.
    Includes index.
    ISBN 1-56158-336-7
    1. Masonry. I. Title.
TH5311.K724 1998
693'.1—dc21   98-22333
                                  CIP

To Betty, my wife and lifelong friend and partner,
who faithfully proofread the manuscript and gave me her invaluable
opinions in the preparation of this book.

ACKNOWLEDGMENTS

I'd like to express my appreciation to the following individuals and companies for their cooperation and technical assistance.

Nancy N. Bailey, my editor

Bon Tool Company

The Brick Institute of America and staff

Delmar Publishers (Photos courtesy Delmar reproduced by permission, *Masonry Skills 3e* by Richard T. Kreh, © Delmar Publishers, Albany, New York, 1990.)

Frederick Brick Works

George Hogan, mason

Betty Jane Kreh, typist and proofreader

Lefty Kreh, photographer

Richard T. Kreh, Jr., photographer

National Concrete Masonry Association

Mark Nunn, Senior Engineer, Brick Institute of America

Prosoco Company, Inc.

Superior Concrete Company

Frank Taylor

Herbert G. Tyeryar, Inc., Concrete Contractors

United Concrete Products

CONTENTS

INTRODUCTION

There have been two principal loves in my life—my family and masonry work. I am very proud that masonry has been a family trade dating back over 300 years. Ever since I was a young boy, I have been associated with the masonry trade, specifically laying brick, concrete block, and stone. I have spent over 40 years working as a mason, teaching the masonry trade in a public school system, and serving as a consultant. Over this span of time, I have built several hundred fireplaces and chimneys. I was taught that masonry is not only a trade but also an art form to take pride in and that my work should always be the best that I can do.

As a masonry teacher, I soon realized that a logical systematic approach is the key to learning and perfecting basic masonry skills. Shortly after I started teaching, I discovered that there was a severe shortage of masonry textbooks that really explain and show the basic fundamentals of

masonry skills. This launched me into a writing career. Since that time I have written a number of successful books and many magazine articles for national and international publishers, including a series of articles for *Fine Homebuilding* magazine.

My ability as a photographer is a valuable asset because I can show often-overlooked but important techniques, such as the placement of one's fingers and hands when holding tools, or closeups showing exactly what I want the reader to see and understand. I have come to the conclusion that masonry books or articles really need to show in detail the specific techniques and methods required to master the masonry skills. I think that this book reflects that!

In this book I have shared with you the information, skills, and techniques that I have learned over a lifetime of working in the masonry trade. My goal was to provide the serious homeowner or person who wants to try his hand in masonry work with up-to-date information and current methods. It is also written for a professional who wants to expand his knowledge and craft and perhaps try some different approaches or challenging masonry projects. I believe the book also will serve as a good reference for anyone who is interested in the masonry trade.

The book is organized in a logical and systematic manner. I start with how to select masonry materials, tools, and equipment required, then move to bond patterns used in brick and block work and how to estimate masonry materials. Next I cover the fundamentals and techniques of laying brick, selected brick projects, how to estimate and lay concrete block, and the fundamentals of concrete work, and end with masonry repair and restoration. I include a chapter on concrete because in my opinion, concrete is an essential part of all brick and concrete block jobs in one way or another.

The projects presented in this book are just a sampling of what one can design and build with brick or block. The possibilities are really endless. Considering the high cost and difficulty of obtaining a professional mason, particularly for small jobs, it is a decided advantage to be able to do your own work. It not only offers you a way to save money on your project but also is an enjoyable pastime and a relaxing form of therapy for the stressful times we live in.

One of the greatest statesmen of this century was Sir Winston Churchill, Prime Minister of England during World War II. He was also an amateur bricklayer who took great pride in showing visitors the brick walls that he built around his country estate. I hope that you derive that same kind of satisfaction from your work and will enjoy this book as much as I have writing it for you.

Chapter 1

MASONRY BUILDING MATERIALS

BRICK

CONCRETE BLOCK

MORTAR MIXES FOR
BRICK AND BLOCK WORK

MISCELLANEOUS
MATERIALS

The first step before building any masonry project is acquiring the needed materials. In this chapter I walk you through a masonry supply house and discuss the various types of materials that you may need. I also cover a lot of dos and don'ts to remember when selecting and buying materials. I have included some suggestions and tips that I have learned the hard way over the years, and they should save you money and frustration in the long run. Let's start out with a little background information on the most popular masonry unit—brick.

BRICK

Brick has been used as a building material for many centuries. The earliest brick were made of clay, formed into shapes, and dried in the sun. These were called *adobe brick* and are still used in many countries throughout the world. As the years passed, someone discovered that if a brick was burned with fire or heat, it would harden and last a lot longer. Who knows how this discovery came about. Someone may have propped up his cooking pot with a brick and stumbled onto this secret. Modern brick are made of clay and shale and are fired in a computer-controlled kiln until hard. Brick are made in a variety of colors and sizes.

My choice for where to buy masonry materials is a building supply house that sells to professional masonry contractors as well as to the general public. Due to a very high volume of sales, its prices will usually be lower than a home center that carries a lot of other items with

Brick panels in a masonry supplier's showroom present many choices of brick as well as different mortar colors and tooled finishes.

masonry materials as a sideline. You will save money, and the selection of masonry materials will be much greater. You can also usually find everything you need at one location, which saves a lot of running around. And the staff at a building supply house can be very helpful in advising you on the amount of materials needed for the job. Chapter 4 provides specific information on estimating brick and mortar, and Chapter 6 covers estimating concrete block and mortar.

Choosing Brick

Before getting started, you should be aware that there are two ways to describe brick size. One is the nominal size, which is the brick size plus the mortar joint, and the other is the actual size—the manufacturer's specified size. The nominal size is most commonly used when ordering brick. (Nominal and actual sizes of masonry units are discussed in more detail in Chapter 4.)

When you visit a supplier, don't be confused by the many types and colors of brick on display. There are really only five major types of brick that you need to be familiar with—the rest are just variations—and you are probably going to select one of these for your job (see the illustration on p. 6). The most frequently used standard brick measures approximately 3⅝ in. x 2¼ in. x 7⅝ in. (nominal 4 in. x 2¼ in. x 8 in.) and weighs about 4 to 4½ pounds. *Norman brick* is the same width and height as a standard brick but is 11⅝ in. long (nominal 12 in.). *Roman brick* is popular for fireplace work, is a much thinner brick, and measures approximately 3⅝ in. x 1⅝ in. x 11⅝ in. (nominal 4 in. x 1⅝ in. x 12 in.). *Engineered brick* (more commonly called *oversize brick* in the trade) measures approximately 3⅝ in. x 2¾ in. x 7⅝ in. (nominal 4 in. x 2¾ in. x 8 in.), and *utility brick* (more commonly called *economy brick* in the trade) measures approximately 3⅝ in. x 3⅝ in. x 11⅝ in. (nominal 4 in. x 3⅝ in. x 12 in.).

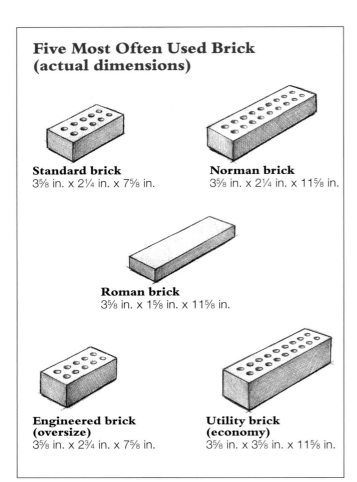

Five Most Often Used Brick (actual dimensions)

Standard brick
3⅝ in. x 2¼ in. x 7⅝ in.

Norman brick
3⅝ in. x 2¼ in. x 11⅝ in.

Roman brick
3⅝ in. x 1⅝ in. x 11⅝ in.

Engineered brick (oversize)
3⅝ in. x 2¾ in. x 7⅝ in.

Utility brick (economy)
3⅝ in. x 3⅝ in. x 11⅝ in.

Brick display panels show a variety of brick along with manufacturers' name and the catalog product number.

It is important to remember that the length of every brick can vary slightly due to shrinkage and expansion from the firing process in the kilns. These tolerances are permitted by the appropriate American Society for Testing Materials (ASTM) standards that govern the manufacture of masonry units.

Another advantage of visiting a masonry supplier is that there usually is a showroom with sample display boards of the various brick laid up in simulated mortar joints on small sections of wall with the joints tooled in popular finishes. You can see exactly how the brick will appear on a finished wall and you can also see the color of the mortar joints. This is important, as some brick look better with a darker or lighter mortar color.

Generally the brick panels are laid half over each other in what is commonly called a half-lap pattern or running bond. Depending on the manufacturer, there will be a catalog order number on the edge or back of the panel, the manufacturer's name, and sometimes a particular name for the brick such as Colonial Pink. Jot down this information so there is no misunderstanding when the brick are delivered. It is a good idea to ask the salesperson when they are available, as some brick are made only in kiln runs at certain times of the year. I also suggest that you ask if additional brick will be available at a later date in case you run out, as some brick are made on a limited basis and then discontinued. If you ever want to add on to your project, finding matching brick could be a big problem.

Characteristics of Brick

Brick are available in a large variety of color and textures. There are smooth face, speckled, matt or scored, sand-finish antique, hand-molded colonial, and so forth. Different tooled mortar joints enhance the various textures of the brick when they are laid in the wall. Selecting the texture of brick depends on an individual's taste, but the texture should fit into the basic surroundings that it will be used in. For example, a sand matt or rough-textured brick would look better in a wooded or natural environment than a smooth brick.

Brick are made solid, with cores (vertical holes), or with frogs (depressed areas in the bottom). A core or frog allows the mortar to lock the brick more securely in the mortar bed joint. A frog should always be laid facing down in the mortar joint. Brick with frogs usually have a slight lip on the top exterior face edge. This lip should always be laid up and to the line, as a straighter wall will be formed if all of the lipped edges are laid up the same way. This lip is particularly evident on colonial reproduction brick and is commonly known as a colonial lip. It is fairly easy to tell the back side of a brick from the front, as many brick have one straighter side due to warped places or kiln marks that may have occurred from lying on their edge in the kilns. If you remember these little tips, it will greatly reduce the possibility of laying brick backward or upside down.

Brick are available with frogs or cores or just solid. Brick with frogs or cores can be locked more securely in the mortar joint.

If you buy brick with holes or frogs, always inquire if solid matching end brick are available for the ends of walls, windowsills, wall caps, and so forth. As a rule they are, but you do have to ask. When selecting brick for flat or paving work, you need to specify this to the salesman, as these brick are made differently from regular solid brick. They should be classified as Grade SW, which means severe weathering, and will cost a little more. (If you want to be sure you are getting a true paving brick, order them as ASTM C902 and Class SX.) Don't learn the hard way—just any solid brick will not hold up for paving work! I also recommend that you buy only a smooth- or sand-surface brick regardless of the size or shape for paving work because they are easier to keep clean than other textured brick. Dirt or mold will collect on the surface of brick otherwise.

How Brick Are Sold

Some years ago I was doing a small brick job for a friend of mine. His wife asked me how much brick cost per dozen. I laughed and told her brick were not sold by the dozen like doughnuts! Brick are generally sold by the piece, the cube, or the thousand lot. If you are buying a truckload of brick, they will probably be less expensive

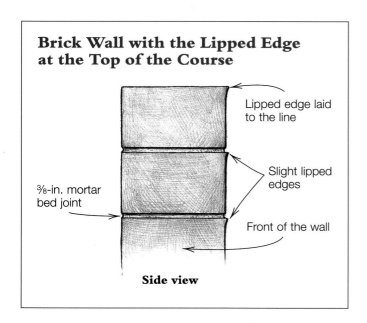

Brick Wall with the Lipped Edge at the Top of the Course

Lipped edge laid to the line

Slight lipped edges

Front of the wall

⅜-in. mortar bed joint

Side view

Cubes of brick are held together with metal bands and weigh more than 2,000 pounds each.

than buying them individually. There is always an extra delivery cost, especially for small lots, which depends on where they are delivered, so ask how much this is. If you are only buying small amounts and have access to a pickup truck, it will probably pay to haul your own.

When buying 500 brick or more, you will find them packaged in cubes held together with metal straps (see the photo at left). The number of brick in a cube varies some depending on their size and how the manufacturer packages them, but generally a cube of standard brick contains approximately 500 to 524 brick. A cube of brick will weigh in excess of one ton (2,000 pounds) and is too heavy to haul in a half-ton pickup truck. It is wiser to make a couple of trips than to risk sitting somewhere along the road broken down.

CONCRETE BLOCK

Concrete block are not new on the building scene but have been around for quite a long time. They are called concrete block because they are made basically of portland cement and some type of aggregate (coarse filler) such as cinders, fine crushed stone, or a lightweight material like fly ash or expanded shale. Concrete block differ from brick in that they are not fired but cured in kilns with high-pressure steam until they are hard. It takes about 4 hours in the kiln to cure a block (com-

Tips on Buying Brick

• Stay away from exotic brick colors and sizes because they will likely be difficult to obtain if you ever need more.

• Have brick delivered on wood pallets, if possible, to keep them off the ground.

• Cover the brick to keep them dry until you're ready to use them.

• Don't cut the metal restraining straps on the brick cubes until you're ready to use them. Wear safety glasses and gloves when cutting the straps.

• If you can't be on the job to receive the brick delivery, put up a sign to the truckdriver to unload the brick close to the area where you will use them. The truck driver will always unload them to his advantage if you don't specify this! Having the brick where you want them

will save a lot of back-breaking labor and will reduce the risk of chipping and breaking if you have to move them around repeatedly.

• Avoid the temptation to buy more brick than you need. You'll have to store them and, believe me, they'll always be in the way. If you expect the supplier to take them back, there will be an expensive handling charge, and it's not worth the hassle.

Assortment of Popular Concrete Block (nominal dimensions)

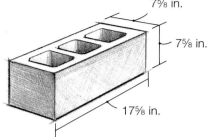

7⅝ in.

7⅝ in.

17⅝ in.

Double corner
8 in. x 8 in. x 18 in.

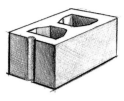

Corner sash

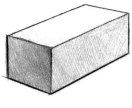

Solid
8 in.

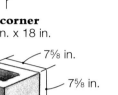

7⅝ in.

7⅝ in.

7⅝ in.

Half corner sash
8 in. x 8 in. x 8 in.

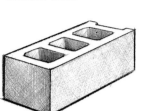

Single bullnose

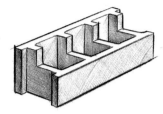

Header

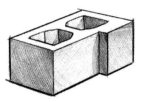

L-corner return

Closed bottom bond beam

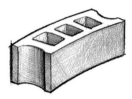

Silo block

Chimney block

Fluted

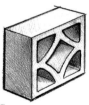

Screen

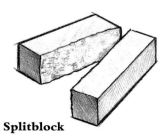

Splitblock

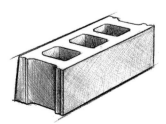

Colonial siding regular
8 in. x 8 in. x 16 in.

Siding combination corner
8 in. x 8 in. x 16 in.

pared to approximately 36 hours for brick). There is very little shrinkage or expansion of block during the manufacturing process, which results in a product of consistent size.

Concrete block are popular among builders and homeowners because they are durable, economical, strong, and easy to lay. A wall can be built of block much faster than brick because they are considerably larger. For example, it would take 12 standard brick to equal the same total area as one 8-in. x 8-in. x 16-in. block. Concrete block can also be easily combined with brick or stone as a backup wall.

Concrete block are readily available in a large variety of sizes and types throughout the country for almost any job. They take paint well and have a good insulation value due to the air space in the hollow cells. Extra insulation can be placed in the cells to make them even more efficient.

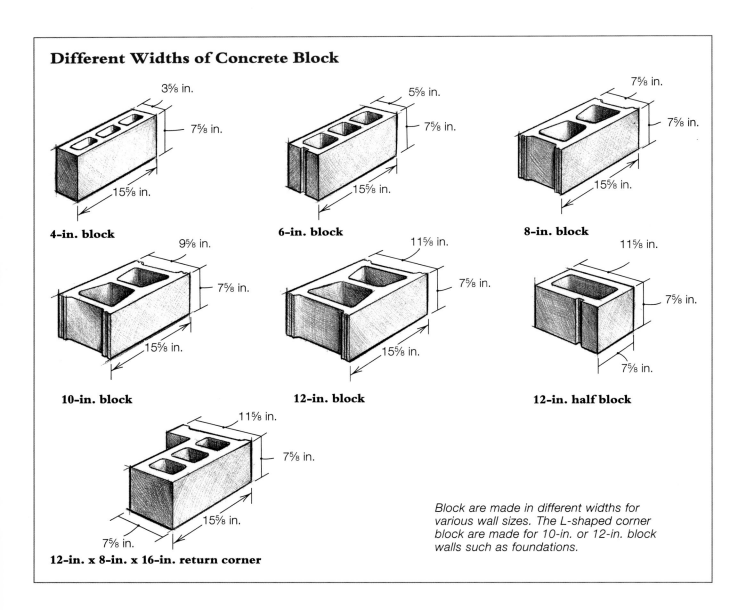

Different Widths of Concrete Block

4-in. block

6-in. block

8-in. block

10-in. block

12-in. block

12-in. half block

12-in. x 8-in. x 16-in. return corner

Block are made in different widths for various wall sizes. The L-shaped corner block are made for 10-in. or 12-in. block walls such as foundations.

All concrete block used to be made with three cells (holes) in them. The trend now is to make two-cell block that line up vertically when laid so that insulation or poured cement grout can be added easily for reinforcement. Another advantage of two-cell block is that they're considerably lighter in weight than three-cell block. It used to be that the majority of concrete block had what was called in the trade *ears,* or projecting ends where the mortar head joint was applied. The majority of block now are square on the ends, eliminating the need to order extra square block for corners, ends of walls, or openings.

Different Types of Concrete Block

There are a variety of sizes and shapes of concrete block available depending on your needs. Like most brick, block are made modular (will fit into a grid of multiples of 4 in.). This is done by design so they will combine and work evenly with other building materials. Almost all concrete block are the same height and length but are made in various widths for different wall thicknesses (see the illustrations on p. 9 and the facing page).

Concrete block are made to be laid with ⅜-in. mortar head (vertical) and bed (horizontal) joints. Therefore, the actual measurements of block are always ⅜ in. less than the figure you commonly hear. For example, an 8-in. standard block really measures 7⅝ in. x 7⅝ in. x 15⅝ in. When the mortar head and bed joints are added, it rounds off to 8 in. x 7⅝ in. x 16 in. This is the nominal size. When you order block, always state the nominal size instead of the fractional size. The length and height are taken for granted. Block range in width from 2-in.-thick slabs used to cap off walls to 12-in. block for foundations or retaining walls.

Heavy or lightweight block

When selecting block, you may be asked by the salesperson if you want heavy or lightweight block. Not too many years ago, all concrete block were made with a lot of cement and fine crushed stone and were very heavy. In the last 20 years, there have been lightweight aggregates manufactured to make lighter block without sacrificing the load-bearing capacity.

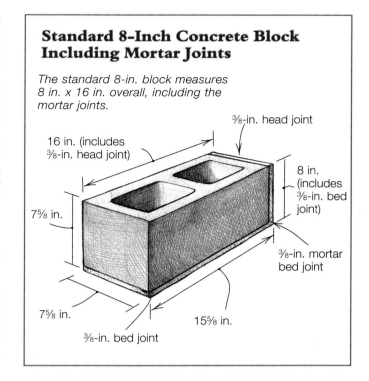

Standard 8-Inch Concrete Block Including Mortar Joints

The standard 8-in. block measures 8 in. x 16 in. overall, including the mortar joints.

⅜-in. head joint

16 in. (includes ⅜-in. head joint)

8 in. (includes ⅜-in. bed joint)

7⅝ in.

⅜-in. mortar bed joint

7⅝ in.

15⅝ in.

⅜-in. bed joint

Since old habits are hard to break, some contractors and masons still prefer to use the heavier block for foundations or retaining walls. Both the heavier and the lighter-weight block comply with building codes. If all things are equal, I would always order the lightweight block for any situation because there is really no difference in their strength or durability. Why wear yourself out lifting a lot of excess weight when you don't have to?

Block for special purposes

There are a variety of special block made for different purposes. *Screen block* are used to enclose privacy areas such as swimming pools, patios, gardens, and so forth, and still permit air to flow through. *Fluted block* have become very popular and have projecting vertical ribs that create a rustic and textured finish. Another popular concrete block product known as *splitblock,* sometimes called *splitrock,* is made of a special coarse aggregate in a 4-in. solid block. The block is then split in half lengthwise at the manufacturer, exposing the rough aggregate surface. Splitblock are laid with the rough aggregate exposed on the face of the finished wall. Generally splitblock are available in white and colored finishes and resemble the texture of a stone wall when built.

There is a special concrete colonial siding block that has been very popular for buildings such as garages. When laid in the wall, it resembles beveled wood siding. The mortar head joints are rubbed flat and the bed joints are tooled. When painted, the wall closely resembles real wood siding unless it is examined up close.

MORTAR MIXES FOR BRICK AND BLOCK WORK

There are two basic kinds of mortar mixes that can be used for brick or block work and flagstone: masonry cement and portland-cement/lime mortar mixes. There are, however, different strengths available depending on specific job requirements. Knowing and selecting the right type can save you a considerable amount of money for your project. All mortar mixes at the masonry supplier come packaged in bags. You should be familiar with and understand the different types available in order to make the best decision for your particular project.

Masonry Cement Mortar Mix

Without any question, masonry cement mortar mix is the most popular type. It is easy to mix and is available in a variety of specific strength designations to fit different job requirements. It comes in approximately 70-pound bags. The general all-purpose masonry cement I recommend is Type N, and unless you specify otherwise, this is the one the building supplier will sell to you. A good way to remember this is to think of Type N as meaning *normal*. Without getting too technical, different strengths of mortars are determined by varying the amount of portland cement in the mix. If you want a stronger mortar mix, ask the salesperson for his recommendation.

I have always considered masonry cement the best all-around buy for the money, and it is by far the mortar of choice for most masons. The rule of thumb when mixing any masonry cement mortar is to use 1 part masonry cement to 3 parts sand and enough water for the stiffness you want. Custom-colored masonry cements are also available that create a colored mortar joint that looks good with certain brick colors or can be used when matching old mortar joints.

Portland-Cement/Lime Mortar Mix

The second type of mortar mix you may want to use is a portland-cement/lime mix. This is composed of Type 1 portland cement, Type S hydrated lime, and sand and requires water to blend it together. A good standard mix is 1 part portland cement to 1 part hydrated lime to 6 parts sand. This is an average-strength mortar mix that has a compressive strength of 750 psi (pounds per square inch) and is classified as Type N mortar. This is more than adequate for most strength requirements for any repair or home construction. Portland-cement/lime mixes can be made stronger or weaker by increasing or decreasing the amount of portland cement.

The principal reason that portland-cement/lime mortars are popular with some masons, builders, and architects is that they do not have any workability additives that could affect the testing strength of the mortar. The lime in the mix increases the ability of the mortar to bond to the masonry unit, which is especially needed in stone masonry because stone are so hard and have a low absorption rate.

For many years you had to buy portland cement and lime separately and mix them together to make mortar. Some masonry suppliers now sell portland cement and lime blended together in one bag. It does cost more to buy it this way, however.

Sand for Mortar

I cannot overemphasize the importance of using clean and properly graded sand for mortar! The grit size of sand for mortar is much smaller and finer than what is used for concrete. The sand should be free of silt, loam, and dirt. When any of these are present in noticeable amounts, they prevent the cement in the mortar from surrounding the sand particles completely, which results in a weaker or poor mix. The presence of loam or mud will also cause the mortar to be very sticky, gummy, and hard to get off the trowel. Mud or silt can cause little holes to show up in the finished cured mortar joints,

This siltation test for impurities clearly shows too much settlement between the sand and the water. This sand would not be suitable for making mortar.

There are a couple of easy ways to tell if the sand is clean. One is to pick up a handful of dry sand and squeeze it into a ball. If it remains in a ball when you open your hand, there is probably too much silt or loam in it. If it releases and falls apart, it should be relatively clean.

The second is a simple siltation test you can perform. Fill a glass jar half full of sand. Add about 3 in. of water. Shake it well, and let it sit overnight. The sand will settle to the bottom and the silt or impurities will rise and settle on top (see the photo at left). If the impurities measure more than ⅛ in. to ¼ in., I would not recommend using it in a mortar mix.

Cost of sand Sand is sold by weight—either by the pound or ton. The more you buy, the less expensive it is. You can obtain the best price for small loads by hauling the sand yourself. Most building supply houses have a minimum delivery charge, which can be very expensive depending on how far they have to go. If you only need a couple of 80- to 100-pound bags, just stick them in the trunk or in the back of a pickup. If you need a half-ton or more, have it delivered at the same time as your other masonry materials. Generally speaking, a half-ton pickup truck should not try to haul more than three-quarters of a ton of sand. I can't count all of the times that I have seen lightweight pickup trucks on the road overloaded or broken down. If you are not careful, the cost of having a small load of sand delivered can wind up costing more than the sand itself!

which is unsightly. It can also cause the finished mortar joints to be somewhat off-color.

How do you avoid this? I always specify washed building sand for masonry work. A reputable masonry supplier stocks this as a matter of habit anyway. You don't have to go to the extremes of telling the supplier the size of the grits in the sand, but just make sure you are getting mortar sand for brick and block work and not a larger-grit concrete sand! Stay away from buying at natural sand pits at bargain prices because they almost always have unwashed sand.

My last tip on buying sand is to have it placed on a clean surface, such as a large piece of plastic or tarp, when it is delivered and not in the grass or mud. You can always make the little ones happy by dumping any leftovers in their sandbox when the job is completed.

MISCELLANEOUS MATERIALS
So far, this chapter has covered the major masonry building materials. Brick and concrete block are the meat and potatoes of masonry work. There are also other masonry materials that you should be familiar with and may need, depending on the project at hand.

Wall Ties and Mortar Joint Reinforcements

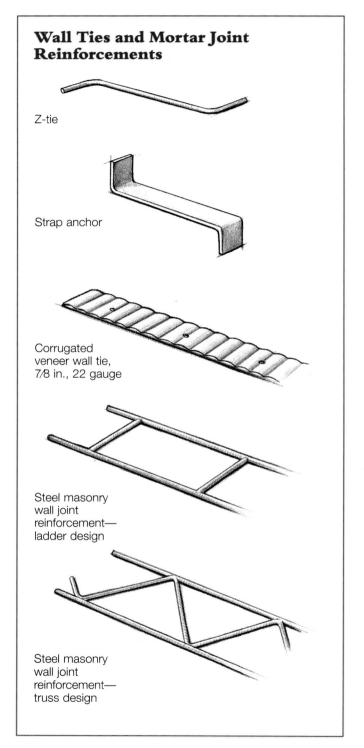

Z-tie

Strap anchor

Corrugated veneer wall tie, 7/8 in., 22 gauge

Steel masonry wall joint reinforcement—ladder design

Steel masonry wall joint reinforcement—truss design

Brick Veneer Wall with a Wall Tie Nailed to the Frame

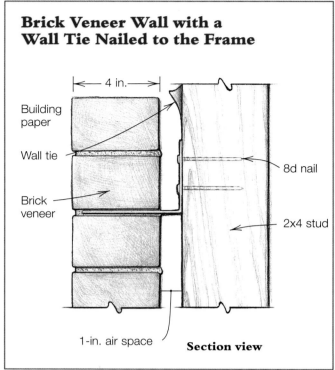

← 4 in. →

Building paper

Wall tie

Brick veneer

8d nail

2x4 stud

1-in. air space

Section view

Metal Wall Ties and Accessories

There will be occasions that you will need metal wall ties or joint reinforcements to tie, bond, or reinforce a masonry project. Some of the ones most often used are shown in the illustration at left.

Corrugated wall tie The corrugated wall tie is most frequently used and is usually made from galvanized metal. It is approximately 8 in. long and 7/8 in. wide. It has a couple of holes in one end so it can be nailed to the wood studs or framing on one end, bent down on the top of the brick, and walled into the mortar bed joints (see the illustration above). Its principal use is to tie brick veneer to the framing of the structure. Corrugated wall ties are relatively inexpensive and are sold individually or in boxes of 1,000. Wall ties are usually installed so that one tie covers an approximate area of 2⅔ sq. ft. Maximum spacings should be 32 in. horizontally and 16 in. vertically.

Z-Tie Bonding Brick to Block in a Cavity Wall

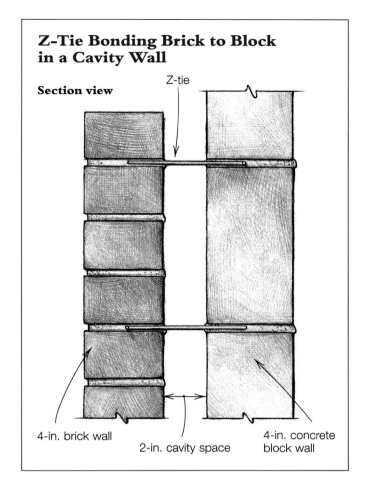

Section view

Z-tie

4-in. brick wall

2-in. cavity space

4-in. concrete block wall

Metal Strap Anchor Tying a Main Block Wall to a Block Partition Wall

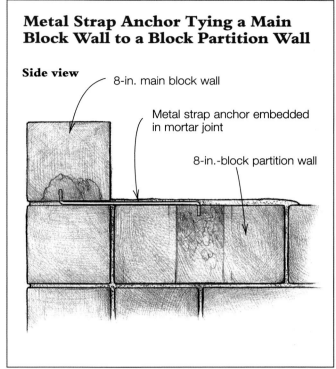

Side view

8-in. main block wall

Metal strap anchor embedded in mortar joint

8-in.-block partition wall

Z-ties Z-ties are made of a heavier metal than wall ties and are used especially for bonding a brick cavity wall to a masonry backing such as concrete block (see the illustration above left).

Strap anchors Strap anchors are heavier yet and made of ¼-in. galvanized steel (see the illustration above right). They are primarily used when bonding a masonry partition or intersecting wall to a main wall.

Steel joint reinforcements Steel masonry wall joint reinforcements are very popular and are used to provide extra strength or to bond two different masonry walls together such as brick backed up with block (see the illustration and left photo on p. 16). The wire is about ³⁄₁₆ in. in diameter, is spot-welded together, and is manufactured in either ladder or truss designs. It is manufactured in different widths for varying wall sizes and comes in standard 10-ft. lengths. I prefer the truss design made by Dur-O-Wall. Joint reinforcement is essential to reinforce masonry retaining walls or to prevent cracking of foundation walls that have earth fill against them. Joint reinforcements are installed every 16 in. vertically.

Fireplace and Chimney Materials

Special materials are needed for fireplace and chimney construction. These include firebrick, clay tile flue linings in various sizes, flue linings that already have the flue ring holes made into them, different size tile flue

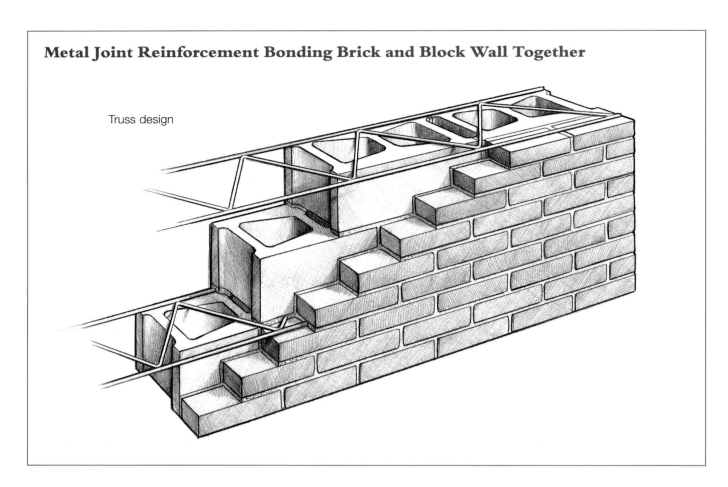

Metal Joint Reinforcement Bonding Brick and Block Wall Together

Truss design

Metal joint reinforcement come in standard 10-ft. lengths.

Use firebrick in fireplaces or to line woodstoves.

A variety of flue liner sizes, ones with preformed thimble holes and thimbles for fireplace and chimney uses, are usually in stock at the masonry supplier.

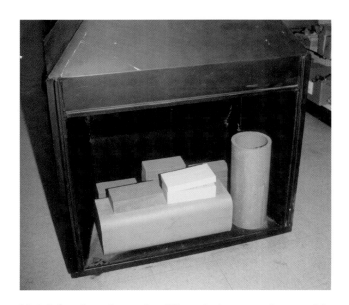

Metal fireplace forms in different sizes can be used in place of traditional masonry fireboxes.

rings for installing woodstoves or furnace connections, angle irons, dampers in different sizes, flashing for the chimney, and metal heat-circulating fireplace units (see the right photo on the facing page and the photos at left and above on this page).

Concrete Steps

Building supply houses also stock precast concrete steps in different sizes (see the top photo on p. 18). They are delivered to the job site and are used to replace old wood or masonry steps. I have found that in many cases I can buy a set of these and set them into place for less money than it would cost to build new brick steps.

Natural Flagstone

Natural flagstone is frequently used in combination with brick and block. Although most masonry suppliers do not always stock regular building stone, they do usually have flagstone. Flagstone is extremely durable and

Precast concrete steps can be a less expensive alternative to brick steps.

Natural flagstone in square pieces cost a little more than irregular pieces.

Crushed stone products and gravel are stocked in quantity for use in retaining walls and patio bases.

beautiful, especially when combined with brick masonry. It is used frequently for outer hearths of fireplaces, patios, walks, and steps. Flagstone can be purchased in random sizes (not squared) or in presawn squares (see the bottom photo on the facing page). The squared pieces cost a little more due to the extra expense of sawing and sizing them.

Flagstone is sold by the square foot. The stone is weighed and then converted to a square-foot figure by the salesperson. The thicker a flagstone is, the more it will cost, as the square-foot allowance is greater. What you are going to use the stone for determines how thick it should be. When buying flagstone for the average job, a good thickness is 1½ in. to 2 in.

For small jobs like driveways, most masonry suppliers will let you pick out your own flagstone. This way you can control the color and thickness that you need. I take a piece of chalk and a steel measuring tape with me to the supplier, mark off the approximate square-foot area

that I want on the driveway, and lay out the stone dry (without mortar). I always allow a couple of extra pieces in case one is broken at the job site. If you run short, you can pick up a few pieces later.

Crushed Stone and Gravel

Crushed stone, stone screening (fine chips), and sand for concrete (see the photo above) are also stocked at good masonry suppliers for use in masonry jobs such as retaining walls and patio bases.

Space does not allow me to discuss all of the masonry building materials that are available. In this chapter, I covered the ones that you would need for an average job. The real advantages of visiting the masonry supplier are that you can see exactly what you're buying, have a better selection of materials, and obtain a better price, plus you can receive helpful answers to any questions you have. As you can see, there is more to buying masonry materials than just picking up the telephone and placing an order!

Chapter 2

MASONRY TOOLS AND EQUIPMENT

HAND TOOLS

MASONRY EQUIPMENT

One of the very first lessons I learned when I started in the masonry trade was that the quality of your work can be improved if you use professional-grade tools. I was fortunate to have this impressed on me at the beginning by my stepfather, who was a master mason with a lot of practical experience in knowing what to look for when selecting tools. The tools needed for masonry work are rather few compared to other building trades such as carpentry, electrical work, and plumbing. In this chapter I share some helpful hints and tips on selecting masonry tools that I have learned over the years.

Number one is to spend your money where it really counts—on the more important tools such as trowels, jointers, and chisels. I always recommend buying brand name tools because they are not only made of a higher grade of materials but also are well balanced and have a longer life. If a brand name tool breaks under normal use or is defective, most dealers I know will replace it free of charge or make a fair adjustment based on wear and the length of time you have owned it. Let's face it—tools such as brushes, squares, chalk boxes, and so forth don't have to be top of the line. You probably already have a number of these lying around the house now. If you are going to economize, do it on these!

Stay away from so-called "bargain" imported tools found in many department stores and home centers. After reading this chapter, you will be equipped with enough know-how to be able to identify them readily.

A standard set of mason's hand tools includes levels, trowels, chiels, jointers, hammers, and squares.

They are really just cheap imitations of good tools and are a waste of money in the long run.

Brand name tools are available at almost all reputable hardware stores or building supply houses. They are also available from catalog companies that specialize in masonry tools. My favorite masonry tool catalog company is the Bon Tool Company. It doesn't carry anything but quality masonry tools at reasonable prices. You can obtain a free catalog by writing to Bon Tool Company, 4430 Gibsonia Road, Gibsonia, Pennsylvania 15504.

The first part of this chapter lists the basic tools required to do brick or block work according to their particular function. They are arranged in the general order that you would normally use them on the job. As you'll soon discover, some of these tools have sharp edges and nasty pinch points, so I've added some safety suggestions throughout the text where they apply.

The selection of masonry tools at a masonry supplier can be overwhelming if you don't know what to look for.

The second half of this chapter covers basic masonry equipment. The list is not large or expensive. My recommendation is to rent costly specialized equipment such as metal scaffolding, a masonry drill, and saws that are used on a limited basis. Keeping in mind the old adage "you only get what you pay for," start off right by sticking with brand name tools.

HAND TOOLS

All masonry tools need to be cleaned or wiped off dry after you are done with them because mortar can build up and harden on them. Any mortar left on them can cause rusting or pitting of the steel surface. Once this pitting occurs, it creates a drag or friction when in contact with mortar and will not allow the mortar to release smoothly. No amount of rubbing with steel wool or grinding will correct this problem, and it usually results in having to replace the tool.

Bricklayer's Trowels

The bricklayer's trowel is used to cut, spread, and handle mortar and is by far the most basic tool of the mason. There are two major styles of bricklayer's trowels. One is the *London pattern* and the other is the *Philadelphia pattern*. Bricklayer's trowels vary in length but still fit one of these two basic designs. It has been my experience that the bricklayer's trowel preferred by masons is the London pattern, which I also use and recommend. It is made in either a wide or narrow width, known as a *diamond heel*. The Philadelphia pattern has more of a square heel and will hold considerably more mortar than the London pattern. I've found that the narrower heel trowel is easier to use without disturbing the line when laying brick, particularly on brick veneer work. I also prefer a smaller trowel because it doesn't hold as much mortar weight and is easier on my wrist. Which you choose is a matter of personal preference.

I recommend buying a medium-length trowel, about 10½ in. to 11 in. long. The length is usually marked on the handle in plain view. This size works well for laying brick or block and is easier to handle than a longer trowel. When loaded with mortar, a trowel that is too large may cause your wrist to ache. Even professional masons have to build up their wrist muscles when they first start out in the trade.

Narrow and wide London pattern bricklayer's trowels are two variations of the most commonly used style.

Quality bricklayer's trowels are made from highly select tempered steel. The blade is drawn out and forged to the shank where it fits into the handle. This one-piece construction is very strong and seldom ever breaks, even with hard use. The cheaper trowels are spot-welded together and will break easily under stress. Handles are made of wood, plastic, or leather. I prefer either the traditional wood or plastic handle. The leather handle is very comfortable, but it will not last nearly as long as the others because trowel handles are used to tap brick into mortar beds.

The angle at which the handle is attached to the blade is known as the *set* of a trowel. Not all trowels have the same set! The reason for the set is that the wrist can become strained and uncomfortable if held in a straight awkward position for a prolonged length of time. Different sets fit some people better than others. This angle also helps prevent dragging your fingers in the mortar joints or against the edge of a brick when spreading mortar on the wall. I can't give you an absolute hard-and-fast rule for choosing this angle. It's really a matter

Choosing the Right Trowel

For most professional masons, choosing a trowel is not taken lightly. There are a few quick on-the-spot tests you can make at the store to determine if the trowel is a quality product and if it is right for you.

First, tap the blade of the trowel gently on a hard surface. If it emits a clear metallic ring instead of a tinny or dead sound, you can be pretty sure that it is tempered steel. Hold it in your hand and see if the handle fits comfortably without grasping tightly. It should feel like it grew there! Next, check the balance of the trowel by allowing it to hang downward on your index finger at the point where the shank meets the handle. The trowel should hang evenly in a vertical position quickly if it is balanced properly (see the photo below left).

The blade of a good bricklayer's trowel should not be too stiff and should flex when spreading or furrowing mortar joints. This flexing movement helps reduce the tension and strain on the fingers and wrists. Check the flex of the blade by holding the trowel by the handle and pushing the point against a hard surface, flexing the blade at the same time. The amount of bending or flexing is a matter of personal choice, but I like the blade to bend at least 1 in. It should spring back easily to its original shape when released (see the photo below right).

Although this may sound like a long, drawn-out process, it really only takes a few minutes to do and will pay off by your selecting a trowel that fits your hand and is a lot more enjoyable to use.

Check the balance of a trowel in this manner. It it's balanced correctly, it will hang evenly.

Test the flex or bend of a trowel.

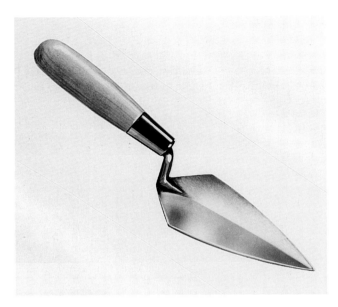

A pointing trowel is useful for getting into mortar joints or holes where a regular trowel won't fit.

Brick Hammers

Hammers used for brick or block work are usually the same general design (see the photo on the facing page). They have a square head on one end and a chisel peen cutting edge on the other. The square end is used for a number of things, including driving nails, striking chisels, breaking block or brick, and tooling a V-groove in mortar head joints.

The cutting edge of the standard brick hammer is on average 1 in. x ¼ in. For cutting concrete block, wider blade hammers are available that are about 1¾ in. to 2 in., but I use the same standard size brick hammer for all of my work. If you are going to own only one brick hammer, my recommendation is to buy an 18-ounce hammer, as it will serve almost any purpose. Handles can be of wood, fiberglass, or metal. Although it is true that the fiberglass handle will resist splitting, I have always used the traditional wood handle. If the wood handle does happen to split or work loose, it can be easily replaced.

of personal preference. My recommendation is to hold the trowel in your hand and choose the set that seems most comfortable to you.

You can expect to pay about $30 for a quality bricklayer's trowel. My two top brands of bricklayer's trowels are W. Rose and Marshalltown.

A smaller version of the bricklayer's trowel is called a *pointing trowel* and is used for pointing up small holes or for places where a regular trowel will not fit (see the photo above). It is invaluable for repair work and pointing up mortar joints in flagstone and stonework. It should be of the same quality as the regular bricklayer's trowel. Pointing trowels are made in different sizes. I prefer the 5-in. x 2½-in. size. A good pointing trowel costs about $11.

Brick hammers are made of tempered forged steel so that they can cut masonry materials easily without undue wear or without the cutting edge breaking. There are also special brick hammers that have a bonded carbide tip edge and cost more than a regular hammer. A good-quality brick hammer costs about $30.

It used to be that the only way you could sharpen a brick hammer was to take it to the local blacksmith and have him heat it in a forge and hammer out the edge. Nowadays it has become very difficult to find a blacksmith. When your brick hammer dulls, I recommend having it sharpened by grinding on a water-cooled wheel. Any good machinist can do this.

I also carry a smaller version of the brick hammer known as a *tile hammer* in my tool kit. It works great when cutting thin pieces of brick, block, and especially clay flue liners, where the shocking power of a normal hammer would break them. A good weight for a tile hammer is about 4 ounces to 6 ounces. It costs about $20.

The weight of a standard brick hammer (*top*) cuts masonry materials easily, while a tile hammer (*bottom*) is better used for thinner materials.

Levels

The level is the most delicate tool the mason uses. It is used to establish a true horizontal or vertical line. Because a level is rather fragile, select a well-made one and treat it with care. The traditional bricklayer's level is 48 in. long and is also called a *plumb rule* in the trade. Levels are available in other sizes such as 42 in., 36 in., 24 in., and 18 in. Most masons I know carry a 48-in. size as well as a 24-in. size for tight places (see the left photo on p. 26).

Levels can be made of metal or seasoned hardwood bound with a metal I-beam frame for extra strength. Many quality wood levels have the edges bound with brass and have metal butt plates on each end. My personal preference is a level with a metal I-beam frame inset with kiln-dried mahogany wood. The EXACT brand level has been my choice for many years. It has tough clear Pyrex-brand vials with air bubbles suspended in a liquid, which remain constant in size in hot or cold weather. The longer levels generally have a hole near the top that allows them to be hung up on a nail when not in use and handholds for easier handling. The smaller ones have grooves in place of handholds.

Most masons carry both 48-in. and 24-in. wood levels.

Use a level to help plumb a brick jamb.

A first-class level will have double vials or sets of bubbles in the level and plumb positions. This allows you to use it vertically and horizontally. It can also be used if one of the bubbles is broken by simply reversing it. A less expensive level with single vials will not allow you to do this, so this is an important feature to consider when buying a level. You can expect to pay about $60 for a good-quality 48-in. level and $50 for a 24-in. level.

Spacing Rules

There are three types of spacing rules available for laying off individual course heights of masonry units: the *course counter brick mason's spacing rule,* the *modular spacing rule,* and a recently developed *oversize scale spacing rule.* The course counter spacing rule and the modular spacing rule are used for the great majority of brick and block work. You will need both for dividing mortar joints when building various brick projects. Their use will be explained in detail in Chapter 4. The oversize scale spacing rule is rather new and is only for larger brick sizes. I don't recommend buying one at first.

When buying any masonry scale rule, select one of the more heavy-duty brands, such as Lufkin or Stanley. The folding ones are made of seasoned hardwood with a standard 6-ft. dimension on one side and masonry scales on the opposite side. A quality rule will have brass joints between each section and on the ends to prolong

Care of a Level

Since a level is a rather delicate tool, don't abuse it. When plumbing or leveling brick or block, never beat on the level to make the wall true. It will not stand this rough treatment for very long before the metal edge bands come loose or the vials or bubbles break! If you do break a vial or bubble, you might be able to send it back to the manufacturer for repair or possibly replacement at a reasonable cost. Another lesson I learned the hard way is never to leave a level in the rear window of a car for a prolonged length of time, as it will warp from the heat of the sun.

I periodically rub the wood on my levels with a light coat of boiled linseed oil on a cloth to preserve the wood and to prevent mortar from sticking to the surface. Boiled linseed oil is available from any hardware store. Be careful not to coat the glass area as it will build up a film and make the bubble hard to see. However, this film can be easily removed with a piece of fine steel wool.

To preserve a wood level, rub it periodically with linseed oil.

its life (see the photo at right). If the joints get too dry, the rule may break at those places. I recommend lubricating the brass joints periodically with a little furniture paste wax rather than with oil. Because a folding rule usually is carried in the hip pocket, the paste wax in the joints will not stain your pants as oil will. The current price for mason's spacing rules is about $18.

Steel Measuring Tapes

The same features of folding scale rules are available in steel measuring tapes. These really come in handy when marking off windows and door frames, and they can hook right on your belt. No tool kit would be complete without a large steel measuring tape. The most popular sizes are the 50-ft. or 100-ft. lengths. I like the 100-ft. tape because it fits most situations without my having to add on measurements. It costs approximately $44.

Chisels

I carry three types of chisels as well as several small ones for trimming (see the top photo on p. 28). This variety will serve just about any situation that you run into. The first one is the *brick set* (also called the *blocking*

A mason's modular spacing rule, scale steel tape, and 100-ft. steel tape are useful for laying off courses of brick or block.

The plugging, mason's, and brick set chisels are essential chisels in the mason's tool kit.

A brick set chisel blade has a beveled side that ensures a clean cut.

chisel). The blade is usually the same width as a brick—about 4 in. One side of the blade is beveled and the other is straight. When making a cut, you should always face the straight side of the blade toward the finished side of the cut. The beveled edge forces the blade down at a slight angle, cutting off any brick irregularities and making a clean cut (see the bottom photo at left). This allows the cut edge of the brick, where the mortar joint is applied, to fit against the adjacent brick neatly. A decent brick set chisel costs about $9.

The second chisel I recommend is a standard cutting type listed in the tool catalog simply as a *mason's chisel*. It is used for everyday cutting out and trimming around masonry work. It is also excellent for cutting concrete block, brick, and stone where the grain lies in layers. The mason's chisel is available in different blade widths, but I would stick with the 2¼-in. x 7½-in. size as an all-around good choice. The cost is about $9.

The third chisel I recommend is called a *plugging chisel* (also known as a *joint chisel*). It has a tapered blade that is very efficient at cutting out mortar joints in old work. The design of the tapered blade allows it to remove the old mortar joint without binding or damaging the face of the brick. It is an absolute necessity for cutting out and repointing old mortar joints. This chisel is made in a variety of sizes. If you are only going to own one, buy the ⅛-in. x 8-in. blade size, as it will fit most situations. You can expect to pay about $8 for this chisel.

Jointing Tools

A *jointer* (also known in the trade as a *striking iron*) is used to finish masonry mortar joints. This tooling or jointing process seals and adds an attractive appearance to the mortar joint. The type of joint finish combined with different textured brick can really enhance the finished look of any masonry wall. It's like the icing on the cake!

A basic set of masonry tools should include five different jointing tools (two of which are shown in the top photo on the facing page). First and foremost is the *convex jointer*, which forms an inverted rounded impression in the mortar joint. Be sure that you specify "convex" when buying or ordering one or you may receive a con-

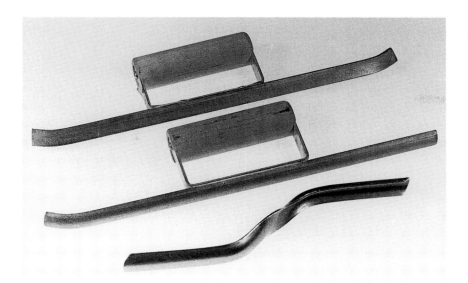

A V-shape sledrunner jointer, convex sledrunner jointer, and pocket-size convex jointer are three must-have masonry tools.

cave jointer, which is used in stonework and forms a beaded protruding joint. In other words, keep in mind that the shape of the tool forms the reverse impression in the mortar joint.

Convex jointers are available as long sledrunner types that form a straight joint in a wall without dipping up or down or as smaller versions that will fit into your hip pocket. I prefer the longer sledrunner type because it is more efficient and quicker to use. The width of the runner can vary. For all-around general use, I recommend the ⅝-in. x 14-in. size. On many professional masonry jobs, architects will specify that only a sledrunner-type jointer be used to ensure the appearance of straighter bed joints. The majority of mortar joints in the United States are tooled with a convex jointer. A convex sledrunner of this size costs about $10.

The second jointer is known as a *V-jointer* and is also made in the form of a sledrunner. It forms an inverted V-impression in the mortar joints. This joint finish is very popular with rough- or matte-finish brick because it provides a rustic appearance and imparts a sense of depth to the joint. It also is a favorite for concrete block mortar joints. I recommend the ¾-in. x 14-in. size for a V-jointer. It costs about $9.

A sledrunner jointer is used to strike mortar joints quickly and efficiently.

The third jointing tool I recommend is called a *grapevine jointer*. The grapevine jointer has a raised bead of steel in the center of the blade, which forms an indented line in a mortar joint. This type of joint finish is specified

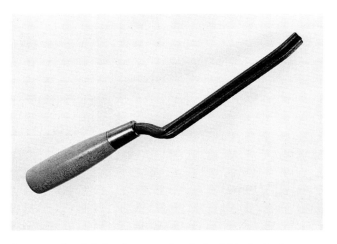

A grapevine jointer has a raised-steel blade to form an indented line in a mortar joint.

Flat slicker jointers—one with a double striking-edge blade and the other with a wood handle—are used to form flat, smooth joints.

A slicker with a ¼-in. blade is best for repointing narrow mortar joints.

frequently for Early American or colonial period brickwork and is used a lot with antique-, matte-, or sand-finish brick. It was and still is very popular in England and Europe. Even though this tool is made in a sled-runner style, I recommend the smaller pocket size because a grapevine joint finish is usually irregular and wavy instead of perfectly straight. I prefer the ½-in. blade width for most work. This tool costs about $9.

The fourth jointer in a basic set of tools is a flat-edge tool called a *slicker* (see the right photo above). It forms a flat or smooth flush joint. Slickers are made in different widths. I carry two different widths to cover most situations—a double-size slicker that is ½ in. on one end and ⅝ in. on the opposite end. If you intend to do any repointing work, then also purchase a slicker with a ¼-in. blade, which is invaluable for small joints or tight places. A good slicker costs only about $5.

The fifth and last jointer that I consider necessary is called a *skatewheel joint raker* (see the bottom photo on the facing page). It is a clever invention that has saved a lot of skinned knuckles! It is used to rake out mortar joints evenly to a predetermined depth. The skatewheel joint raker has two skatewheels and a handle with a hole in the center to insert a nail to rake out the joints.

End View Profiles of Mortar Joint Finishes

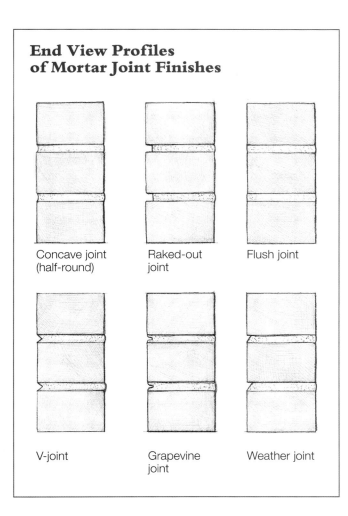

Concave joint (half-round)

Raked-out joint

Flush joint

V-joint

Grapevine joint

Weather joint

Raking out mortar joints with skatewheel joint raker saves lots of skinned knuckles.

It works like this: Insert the nail in the hole located between the two wheels to the depth you want it to rake out below the bottom edge of the wheels. Tighten the thumbscrew that holds the nail in place. Use a case-hardened or tempered masonry nail for long wear. The wheel rides on the edges of the brick adjoining the joint, and the nail rakes out the joint to the preset depth perfectly (see the photo above). Lubricate the wheels periodically with some light oil to keep them working smoothly. I recommend that you never rake out a mortar joint any deeper than ½ in. to prevent leaking. A raker costs about $10.

Squares, Chalk Boxes, and Brushes

I carry a regular 2-ft. framing square for most squaring needs (see the top photo on p. 32) and a smaller 12-in. one for squaring cuts on brick or block. I also find that a smaller carpenter's try square comes in handy. A 2-ft. square costs about $8, and a 12-in. square costs about $4. A carpenter's try square runs about $22.

A chalk box (shown in the center photo on p. 32) is a necessity for laying out wall lines. Traditionally, on the job, tradesmen use blue chalk and building engineers

A skatewheel joint raker uses an adjustable nail to rake out joints to a predetermined depth.

A 2-ft. framing square is useful for most squaring needs.

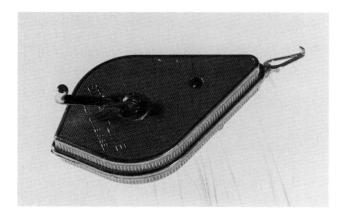

A chalk box is used to lay out wall lines.

Use a bricklayer's brush after the mortar joints have been tooled and have dried enough not to smear.

use red chalk. You can buy a good quality chalk box for about $10 and a plastic squeeze bottle of chalk for about $2.

After the mortar joints have been tooled and are dry enough so as not to smear, they need to be brushed. I recommend buying a 13-in., medium-soft, long-handle stove brush (shown in the bottom photo at left), which will save you from skinning your knuckles on the masonry work. It only costs about $5.

Mason's Line

As far as I am concerned, the only line that should be used as a guide to lay masonry units to is nylon line (shown in the photo below). Nylon line will not rot or break down from moisture or the sun. It is very strong and can be pulled tightly with little sagging, which is important for maintaining a level wall. Nylon line is available twisted or braided and in a variety of colors. I recommend buying No. 18-diameter braided line either in white or yellow so it is easy to see. I prefer a 500-ft. roll of line. The cost should be about $10.50 per roll. To prevent nylon line from unraveling, seal the ends with the flame of a match.

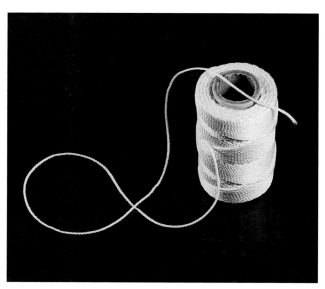

Mason's nylon line is very strong, is available twisted or braided, and doesn't rot.

Line Pins, Line Blocks, and Trigs

After the corner or leads are built, brick or block courses are laid to a line stretched tightly between the corners. The line can be attached to a nail on one end, stretched tightly, and wrapped around a steel *line pin* driven in the mortar head joint on the opposite end. The line can also be held in position with a wood or plastic *line block* stretched tightly against the end of each corner and held by tension (see the photo at right). On longer walls, a thin metal clip can be snapped on the line to keep it from sagging in the center of the wall. This metal clip is called a *trig* and is placed on top of a prelaid brick and held in position with another brick on top of it. The uses of these tools are explained further in Chapter 5.

Many masonry suppliers and manufacturers give line pins, line blocks, and trigs as freebies and advertise their firm's name on them. Check this out before buying any. If you have to buy them, the cost will be about $4 for a line pin, $1 per set of line blocks, and 25¢ for a trig.

A line block holds a line in position, stretched tightly between two corners.

Brick courses are laid to lines with the help of a line block, a trig, or a line pin.

Utility wire cutters are essential—and easier on your hands than cutting pliers.

Utility Wire Cutters

If you are going to use any metal joint reinforcement wire in your masonry work, it is pretty hard to get along without a pair of utility wire cutters (shown in the photo above). Most professional masons keep one of these in their tool bag. A heavy-duty pair of cutting pliers will work in a pinch, but they are tough on your hands. You can expect to pay about $44 for a decent wire cutter.

Tool Bag

After you have collected your tools, you need some type of tool bag or container to keep them in. Many professional masons carry their tools in a traditional canvas bag reinforced with leather on the bottom handles, which is expensive. You don't need this; any heavy-duty canvas, all-purpose tote bag with handles will serve the purpose. You can readily buy one of these for between $35 and $40. Metal tool boxes don't work out well, as there is never enough room in them and it always seems that the tool you need is at the very bottom of the box! If you really want to cut costs, hunt up that old gym bag you had in high school or use a plastic 5-gallon bucket. And don't forget to include a pair of safety glasses or goggles and first-aid kit in your tool kit.

MASONRY EQUIPMENT

The masonry equipment needed is relatively minimal compared to the number of hand tools. In masonry work, there is always some specialized equipment such as saws and drills that you only need now and then. I recommend renting this equipment when you need it. Following is a list of masonry equipment that will make your life a lot easier when doing brick or block work.

Wheelbarrows

You can't do without one of these. There are a lot of lightweight garden-cart types of wheelbarrows on the market. They will not hold up long for mixing mortar or concrete or for moving masonry materials. I recommend investing in a heavy-duty contractor's 5¾-cubic-ft. size. It is made from high-density polyethylene or tough heavy steel. It has seasoned wood handles and 12-ply tires, which will really hold up well under a load. If you expect it to last, make sure that the bed is kept good and clean by washing it out with water after mixing mortar or concrete in it. I have a steel wheelbarrow that I have been using for 20 years and it is still in good shape. A wheelbarrow of this type will cost about $100, but with good care it will last forever.

A wheelbarrow, mortar hoe, and square- and round-point shovels enable you to mix batches of mortar on the spot.

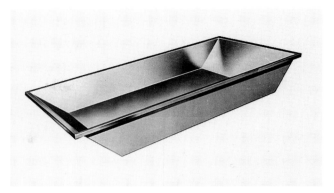

A steel mixing box is more durable than one made of wood.

Mixing Tools

I use two different types of shovels—a regular round-point shovel and a square-point shovel—for cleaning up around the edges of the mixing box and for transferring mortar to the wheelbarrow. I add mortar ingredients to the mud box or wheelbarrow with the round-point shovel because it is more accurate. A *mortar hoe* has two holes in the blade to allow the materials to pass through when mixing mortar. I recommend buying a mortar hoe with a tough plastic handle as it won't rot out or break

prematurely. I also strongly advise buying the smaller size hoe—the bigger one is a real killer. Expect to pay about $25 for each one of these.

Mortar Mixing Box

If you intend to mix a lot of mortar with a hoe, you may want to consider a *mortar mixing box* (see the photo above). It comes in a variety of sizes. A big one will work you to death and is hard on the back. My recommendation is to buy a smaller steel box that measures 48 in. long x 14 in. wide x 11 in. deep. It costs about $75. I have tried without much luck to make wood mixing boxes, but they always come apart and don't hold up for very long.

Mortarboard

Masons use a *mortarboard* or *mortar pan* to hold the mortar and work off of after the mortar has been mixed. Most professional masons prefer the mortar pan because it holds more mortar, lasts a lot longer than the board, and takes a lot of abuse. It is made of steel or rigid plastic and measures about 29 in. x 29 in. x 6½ in. deep. A steel mortar pan costs about $39.

However, all you really need is a wood mortarboard that can be made at home from scraps of leftover lumber or plywood. A mortarboard should be about 30 in. x 30 in. with two 2x4 runners fastened to the bottom to keep it off the ground (see the illustration on p. 36). The best

Wood Mortarboard

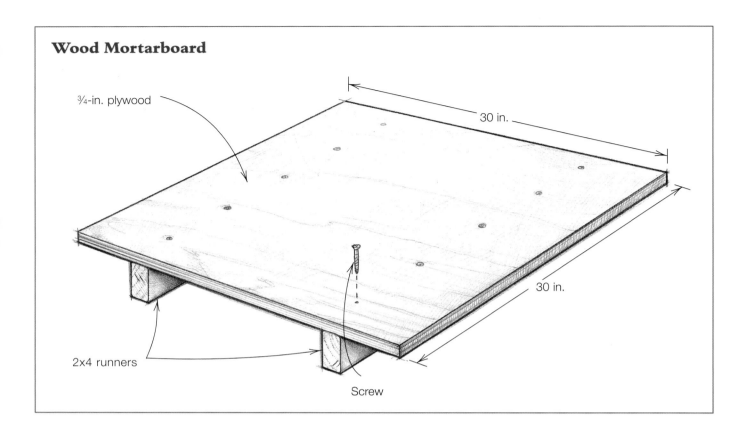

¾-in. plywood

30 in.

30 in.

2x4 runners

Screw

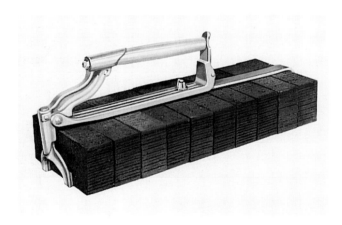

Brick tongs are adjustable to fit different brick sizes and can pick up 10 standard brick at one time.

material to use for the top is a piece of ¾-in. exterior or treated plywood. I recommend attaching the 2x4 runners with screws because nails will work loose after a period of time and are a real nuisance when picking mortar up off the board with a trowel.

Brick Tongs

Brick tongs (or *brick carriers*), shown in the photo at left, are used to pick up a number of brick at one time. They save a lot of time and wear and tear on the fingers. They are adjustable to fit various brick sizes and can handle up to 10 standard brick at one time. They work great for stacking brick and reduce chipping the faces of the brick. Brick tongs is one piece of small equipment that is a must to own if you are handling any amount of brick. The cost of good-quality brick tongs is about $20.

Utility Cement Mixer

Although having a cement mixer is not an absolute necessity, I can assure you that this is one piece of masonry equipment I would not be without. The larger

A concrete mixer isn't a necessity, but it sure beats mixing mortar by hand.

professional mixers can cost well over $1,000. You can purchase one of the smaller utility mixers that mixes mortar as well as concrete for about $250, which includes the electric motor. I recommend buying the 3½-cubic-ft., drum-type mixer. The best place that I have found to get one of these is a farm supply center, as they sell tons of them to farmers for mixing concrete or feed. With proper care and cleaning, it will last a long time.

Incidental Equipment

A length of 50-ft. garden hose with a nozzle attached and several 5-gallon buckets round out the minor equipment list. I like the 5-gallon white plastic buckets that spackling and industrial cleaners come in. You can find these just about anywhere; they're usually free for the asking.

In this chapter, my principal goal was to advise you about the tools and equipment you need to do masonry work without wasting a lot of money. I suggested some helpful tips to follow when selecting these items. In all trades, there are always a lot of gimmicks and gizmos around, which are really not essential. If you stick with the tools I have listed, you'll discover that you are prepared for almost any situation. As I stated in the beginning, stay with the brand names and rent the more expensive equipment since you won't need them all of the time.

My final comment concerns safety: Remember that it is not the tools or equipment that hurt you but the way you use them.

Chapter 3

BOND PATTERNS AND WALL TYPES

BONDS

TYPES OF BRICK
AND BLOCK WALLS

BRICK AND BLOCK PROJECTS
FOR SPECIAL PURPOSES

In the previous two chapters I described materials, tools, and equipment needed to do masonry work. In this chapter I explain the different bond patterns used for brick and concrete block work, different types of walls you can build, and some interesting projects. It doesn't make any difference what type of bond, or pattern, you select for a wall; however, one critical requirement should always remain the same—the wall must be structurally sound.

BONDS

When I think of the word "bond," I think of three different meanings. One is the adhesion of the mortar to the brick, as the mortar has to bond, or hold, the wall together. The second is the interlocking or overlapping of the individual brick or block in the wall, which divides and distributes any load imposed on it. This is called *structural bonding* and causes the entire wall to act as a unified mass. The third is the variety of brickwork patterns or designs, which are creative and pleasing to the eye. These patterns or designs are also called *bonds*. Some bonds are very simple and others are complex. Regardless of the type of bond used, it is important always to maintain the pattern throughout the project or the end result will be very disappointing.

To create bonds, brick are laid in different positions that have been assigned special trade names recognized by masons, architects, and builders everywhere. (See the illustration on the facing page.) You should familiarize yourself with these terms early on, as they relate to the

A brick herringbone panel wall can be a real eye-catcher.

Names of Brick Positions in Brick Walls

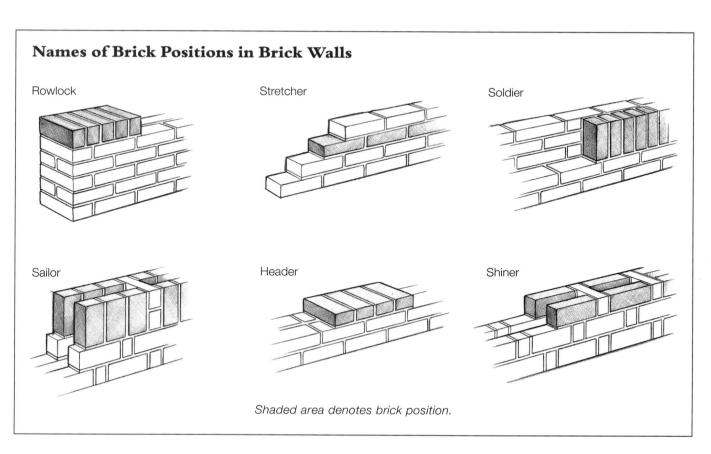

Rowlock

Stretcher

Soldier

Sailor

Header

Shiner

Shaded area denotes brick position.

This is an example of how to use brick headers to corbel out under a picture window.

position of the brick in the wall and will be referred to many times throughout this book. A bond can be laid in the same position repeated over and over or in a combination of different brick positions. Bonds or patterns can also be intermixed in the same project, which is common in flat or paving work. The key is always to pay particular attention to the way a bond is started because generally that is the pattern that should be repeated. There are really not all that many different bond patterns for brickwork.

Running Bond

The simplest of all bond patterns is the *running bond* (also known as the *all-stretcher bond*). This is the one you see most frequently. This bond consists of all full brick (stretchers) with the exception of pieces that have to be cut at windows, doors, or other openings to maintain the pattern. There are two different versions or forms of the running bond—the *half-lap* and the *one-third lap*.

The half-lap consists of each brick laid half over the one beneath it (see the photo and illustration on the facing page). Because there are no header brick to bond or tie the wall to the backing, metal wall ties of some type are used to accomplish this. A half-lap bond is generally used with standard 8-in. brick. Since the end of an 8-in. brick is 4 in. wide, each time it returns on the corner over the brick below, it creates a half-lap without any cutting.

The one-third lap running bond is generally used when laying a 12-in. Norman brick. In this bond pattern, shown in the illustration on p. 42, the 4-in. end of the

Half-Lap Running Bond

This brick privacy wall is laid in the half-lap running bond.

One-Third Lap Running Bond

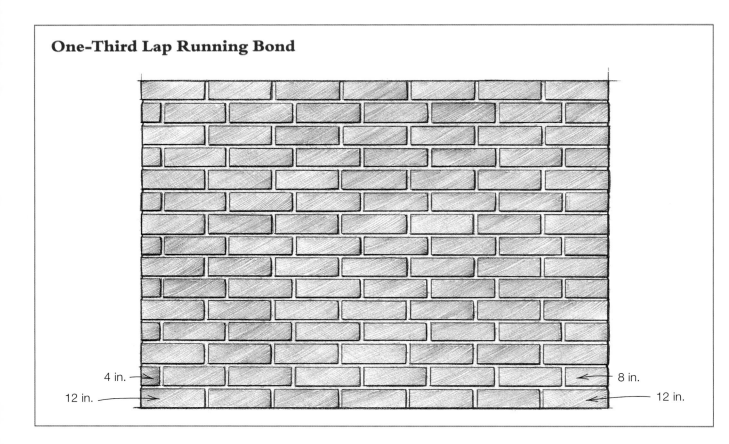

4 in.

8 in.

12 in.

12 in.

Norman brick creates a one-third lap when it returns on the corner because the stretcher brick beneath it is 12 in. long. The Norman brick is popular for fireplace work and in a lot of commercial buildings such as shopping centers.

Because the running bond is a simple pattern, it is by far the most popular of all brick bonds, especially for brick veneer homes. Most brick, regardless of their length, are usually 4 in. wide and that works out perfectly for brick veneering a wood frame building such as the typical house. Due to the simplicity of the pattern, the running bond is very popular with builders and is a high-production bond for the bricklayer to build. It is also the easiest bond pattern to match if an addition is added later to the structure.

Common Bond with Full Headers

The *common bond,* also called the *American bond,* is used for solid masonry walls. It's a variation of the running bond with the difference being that it has a course of brick headers at regular intervals. Generally, a common-bond brick wall is backed up with either 4-in. or 8-in. concrete block to a certain height. The brick header is then laid crosswise in mortar and also rests on the concrete block backup, tying the wall together.

Headers usually occur at either the fifth, sixth, or seventh course of brick, depending on how they match up with the height of the block backing. A good example of a common bond would be a brick and block retaining wall to hold back earth. A piece of brick cut to 6 in. is called a *three-quarter,* and it is required on each end of the header course, either on or against the corner or end to create the 2-in. lap over the brick below. Before brick veneer houses became so popular, walls were made of solid masonry, and the common bond was the bond of choice. It is still used in a lot of renovation work.

Common Bond Brick Wall with Full Header

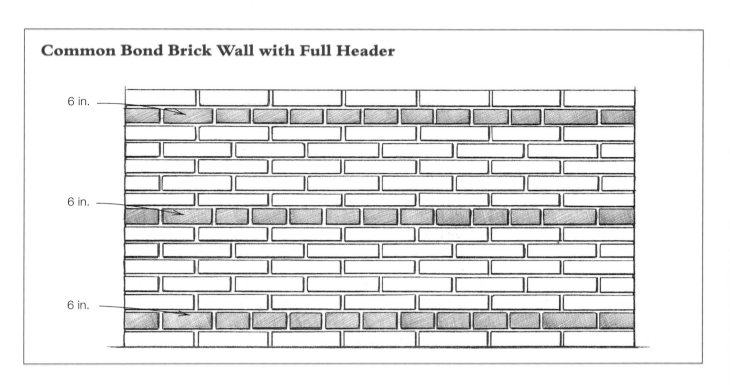

6 in.

6 in.

6 in.

Common Bond Brick Wall Backed Up with Concrete Block

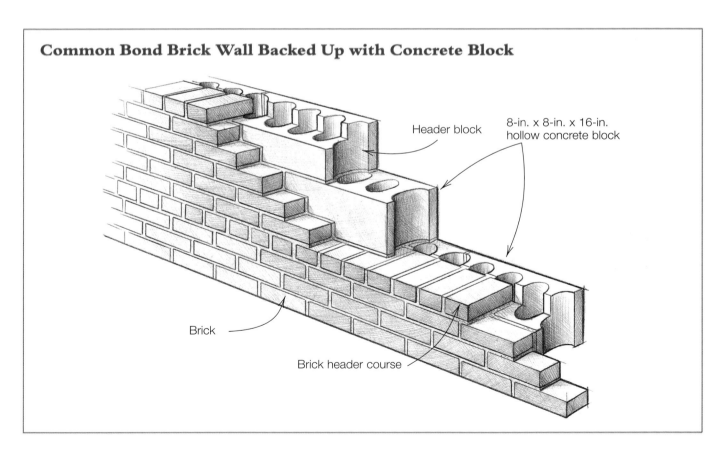

Header block

8-in. x 8-in. x 16-in.
hollow concrete block

Brick

Brick header course

Common Bond Brick Wall with Flemish Headers

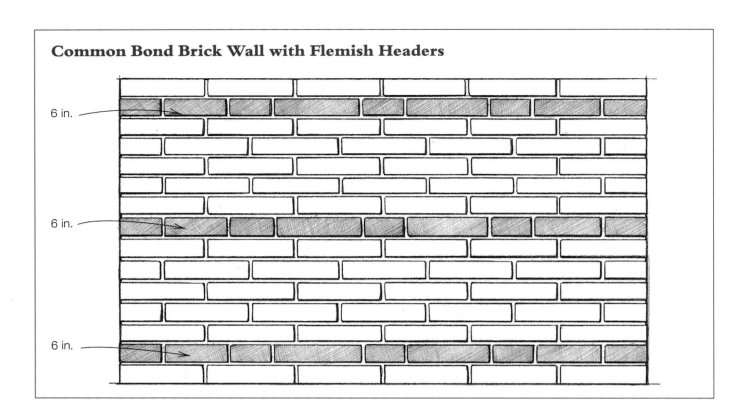

6 in.

6 in.

6 in.

Common Bond Brick Corner with Flemish Headers

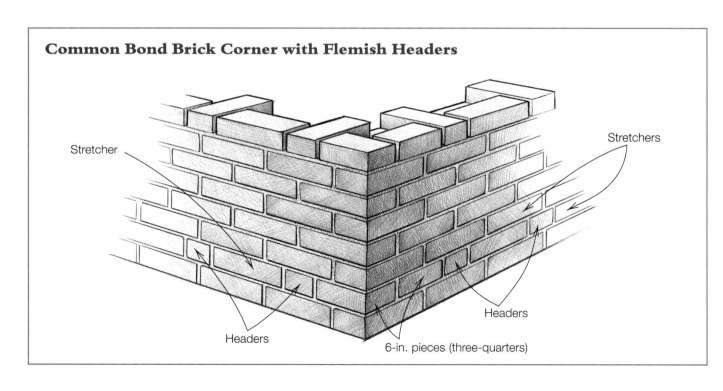

Stretcher

Stretchers

Headers

Headers

6-in. pieces (three-quarters)

Common Bond with Flemish Headers

A common bond brick wall can be varied on the header course by laying what is called a *Flemish header*. The difference is that the header course is a header alternated with a stretcher brick. The full header and the Flemish header together create a strong tie course. In all cases, the headers should be centered over either a stretcher brick or a head joint below. Mortar head joints should never line up one over the other, except in a few special pattern bonds where joint reinforcement is used (see "Stack Bond" on p. 46).

Flemish Bond

The *Flemish bond* originated in England and was very popular when Colonial Williamsburg, Virginia, was built. It is considered to be one of the most beautiful pattern bonds in brickwork. Each course of brick consists of alternate stretchers and headers, with each header centered over the stretcher or head joint below it. The headers located on every course are in a plumb vertical line with each other. Because the wall is tied on every course, a Flemish bond wall is usually built of all brick including the backing. If a Flemish bond is to be a single-thickness wall, such as in brick veneer, then all of the headers must be cut in half to accommodate a 4-in. wall thickness, and wall ties must be used to tie the brick to the framing.

There are two different methods of starting a Flemish bond from the corner or end. They are known as the *Dutch corner*, in which a three-quarter (6-in.) piece is used, and the *English corner*, in which a 2-in. piece of brick (called a *plug* in the trade) is used. The Dutch corner is the more modern version, and the English corner is the more traditional method. If you are attempting to duplicate the original Flemish bond, I recommend using

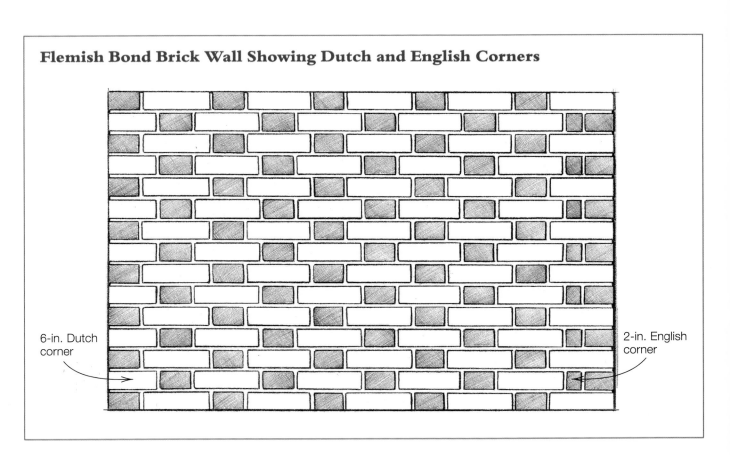

Flemish Bond Brick Wall Showing Dutch and English Corners

6-in. Dutch corner

2-in. English corner

This Flemish bond brick wall with a circular window can be found in a historic church in Williamsburg, Virginia.

the English corner, but it is a matter of personal choice. Many builders of fine homes and buildings, particularly banks, government buildings, libraries, and educational institutions, prefer the Flemish bond.

Garden-Wall Bonds

An adaptation of the Flemish bond, the *garden-wall bond* is popular for brick walls that enclose patios, courtyards, or estates. There are two different versions of the garden wall bond. A bond that has two stretchers and a header alternating on the same course is known as a *double-stretcher garden-wall bond*. A garden wall bond that has three stretchers and a header alternating on the same course is known as a *triple-stretcher garden-wall bond*. Both of these bonds create a diamond pattern. The triple-stretcher garden-wall bond causes the diamond pattern to open up or appear larger.

The bond can be highlighted by combining dark header and stretcher brick to emphasize the pattern. These darker brick are available from a brick supplier. The brick that form the diamond pattern can also be projected or recessed to highlight the pattern. This type of bond requires strict adherence to the pattern development and does take extra time to build. Every brick must be laid properly in its assigned position or the pattern will be ruined. If you want to build a brick panel or wall that is different and that you can brag a little about when it is completed, this is a good one to try. It will be the envy of the neighborhood when you have completed it.

Stack Bond

In the *stack bond*, there is no overlapping of the brick or block. Each masonry unit lines up vertically with the one beneath it, and the mortar head joints are in a

Double-Stretcher Garden-Wall Bond

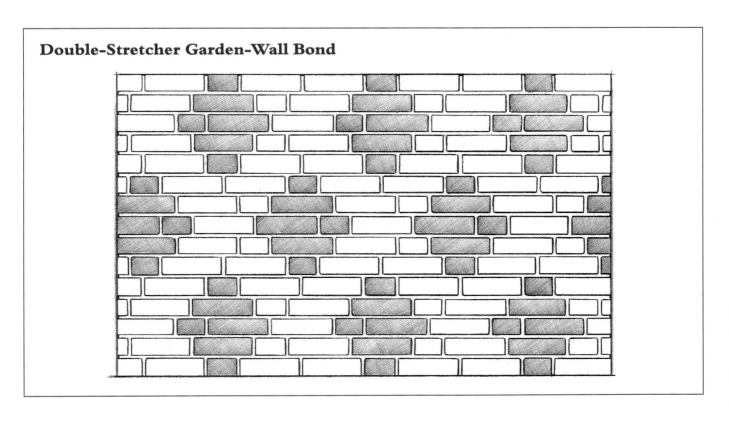

Triple-Stretcher Garden-Wall Bond

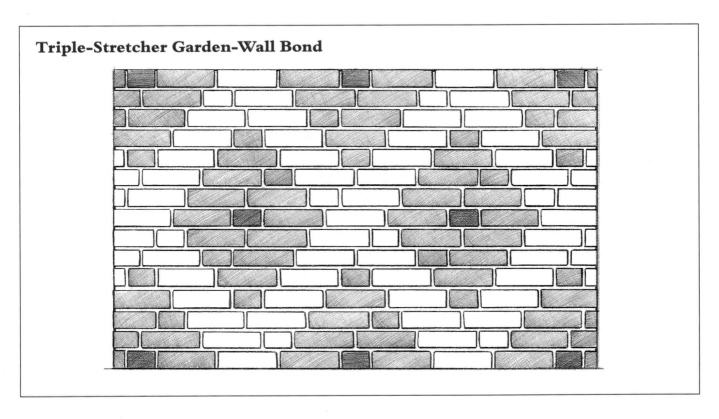

Stack Bond Brick Wall

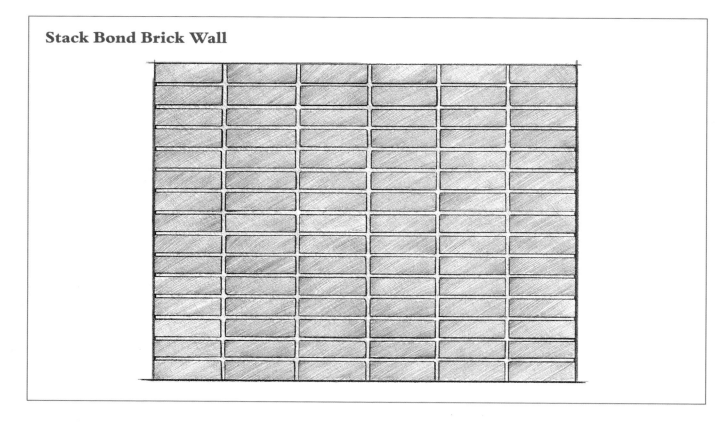

straight vertical line. Although this is a popular pattern for brick and block, it is not the easiest to build. Since all brick are not exactly the same length, you have to pick and choose among them to maintain the plumb vertical head joints. Concrete block work better than brick for stack bond walls because they are steam-cured and are more accurate in length. However, this pattern can be achieved with clay brick units that adhere to strict dimensional tolerances, such as modular brick.

One of the major problems with the stack bond is that because all of the head joints are in a plumb vertical line, they are subject to hairline cracks from contraction and expansion, much more so than a regular bonded brick or block wall. Metal wall ties and joint reinforcements are an absolute must for a wall built with the stack bond and are recommended every 16 in. vertically or every two courses of block high for best results.

English Bond

The *English bond* is made up of alternate courses of headers and stretchers. Don't confuse this with the Flemish bond previously discussed, which has a stretcher and header alternating on the same course. In the English bond, the headers are centered on the stretchers and mortar head joints in all courses below it. The English bond is started from the corners or ends of the wall with a 2-in. piece of brick (the English corner) or with a three-quarter piece (the Dutch corner).

The English bond originated in England as its name suggests. American historical colonial cities such as Williamsburg have many fine examples of this bond. One of its unique features is that it appears to have a repeating diamond pattern. For this reason, it is sometimes called a *diamond bond* by many masons.

English Bond Brick Wall

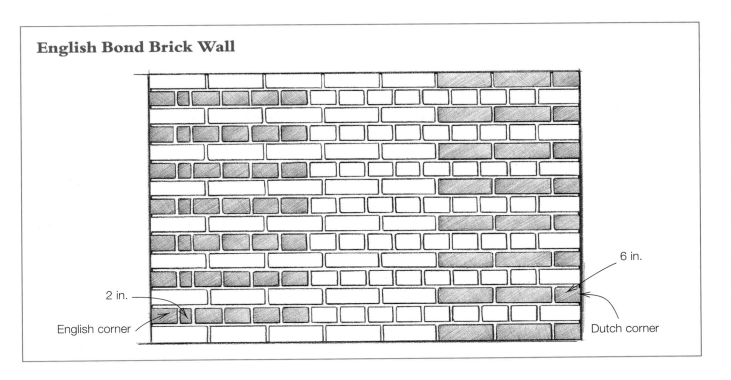

2 in.

English corner

6 in.

Dutch corner

A beautiful brick home in colonial Williamsburg is laid up in the English bond.

Paving Bonds

Brick paving can be laid in a variety of patterns for walks, patios, or driveways. It is easier to install paving bonds than brick in a regular vertical wall because everything is on a relatively flat plane and won't shift as much when you are laying it. As stated in Chapter 2, only solid brick should be used for paving bonds. They can be laid flat or on their edges. A job will take considerably more brick if laid on edge than laid flat. When estimating brick to be laid flat, the rule of thumb I follow is to figure on using one-third less brick laid flat than for brick laid on edge. Brick paving can be laid dry in sand or in mortar.

There are four basic paving bonds; the others are combinations or variations. They are the *basket weave,* the *herringbone,* the *running bond,* and the *stack bond.* The basket weave pattern consists of two full brick laid flat (shiner position) in the same direction and then two brick laid flat in the opposite direction. This arrangement is repeated throughout the paving. The basket weave pattern is very popular for walks and patios and is

Popular Brick Paving Bonds

Basket weave

Herringbone

Running bond

Variation of basket weave

Variation of basket weave

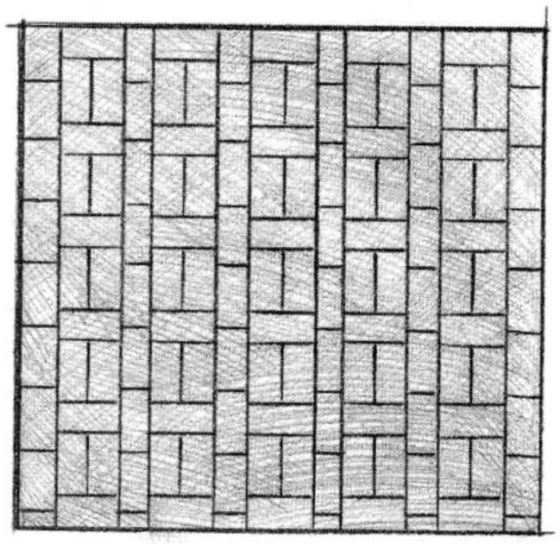

Variation of basket weave

Variation of basket weave

One-quarter running bond

Running and stack bond mixed

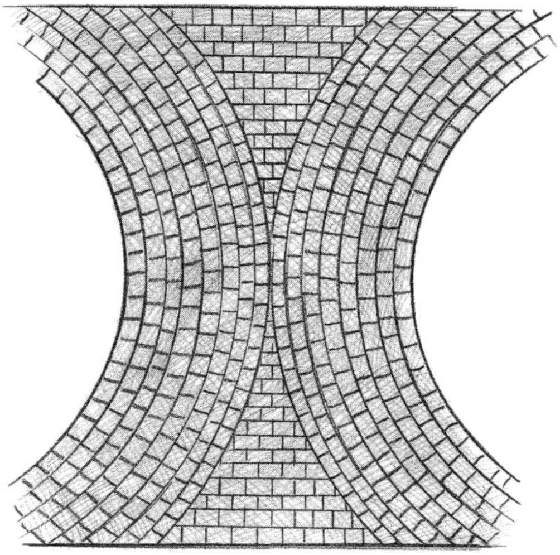

Circular and running bond mixed

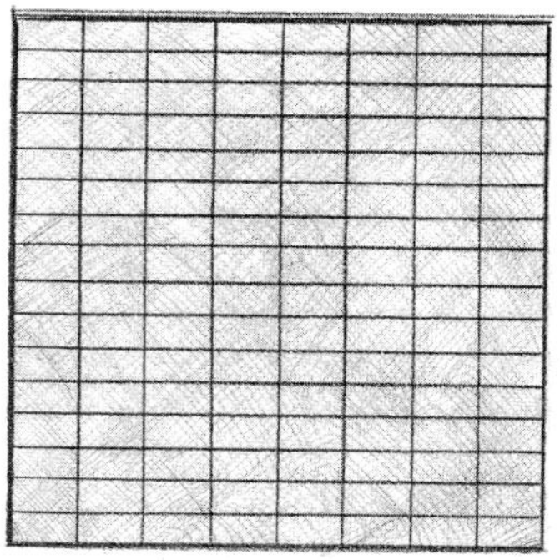

Stack bond

This brick basket weave walk is laid mortarless in a dry bed of sand.

relatively easy to lay. The main thing you want to watch out for is matching up brick of approximately the same length next to each other so there is not too much variation in the joints.

The herringbone pattern is formed by laying each end of a standard nominal 8-in. brick at a right angle to the one next to it in an L-shape. It is a great bond for resisting movement or shifting, as the brick are locked together more tightly than in other bonds. For best results, a herringbone bond should have some type of a bordered edge to restrain it. Generally, the herringbone brick pattern is laid on a 45-degree angle off the border.

The running bond pattern is the same as discussed for walls, with the exception that in paving work, the brick are usually laid in a flat shiner position. For best results, I always recommend a border, preferably of brick, laid in another position to restrain the paving. Treated wood 2x4s also will do a good job.

My two favorite paving bonds for brick are the basket weave and herringbone bond. My last choice is the stack bond. If a stack bond is laid in mortar for paving, modular brick should be used.

A brick walk laid in herringbone bond resists cracking because the brick are locked together tightly.

TYPES OF BRICK AND BLOCK WALLS

There are a variety of different brick and block walls that can be built. The function, need, and requirements of the wall dictate the ultimate design. I always think about factors such as blending into the surroundings, ensuring strength, and creating an attractive project. There are six general classifications of wall types that you should be familiar with because just about any masonry project will fit into one of these forms or another.

Solid Brick Walls

The oldest type of wall is the solid brick wall, which is built entirely of brick regardless of its width. Brick headers, metal wall ties, or joint reinforcements are used to bond or tie the wall together at spaced intervals. Not too many years ago, before the advent of concrete block, which reduced the cost of solid brick walls, all brick walls were built this way. Solid brick walls are still used a lot for garden walls, patios, and masonry structures around the home.

Brick and Block Walls

Brick walls backed up with concrete block (also called *composite walls*) are very popular for garages, foundations above grade, and storage and commercial buildings. The lower cost and paintability of concrete block when combined with brick produce fireproof masonry walls with low maintenance cost. The brick and block can be bonded together on the sixth or seventh course with either brick headers, metal wall ties, or joint reinforcements to form a strong wall.

Brick Veneer Walls

A brick veneer wall consists of a brick facing built up in front of some other material, such as a stud wall in house construction. Brick veneer is by far the most popular type of brickwork used in brick homes today. Insulation of some type is usually installed between the wood studs in the wall behind the brick veneer to make

---

Solid Brick Wall

16 in.

Metal Z-ties can be installed every 16 in. vertically to bond or tie the solid brick wall together.

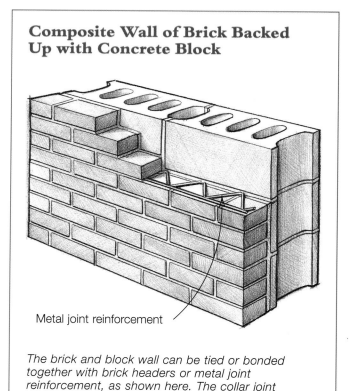

Composite Wall of Brick Backed Up with Concrete Block

Metal joint reinforcement

The brick and block wall can be tied or bonded together with brick headers or metal joint reinforcement, as shown here. The collar joint between brick and block is filled with mortar.

Brick Veneer Wall Built in Front of Wood Framing

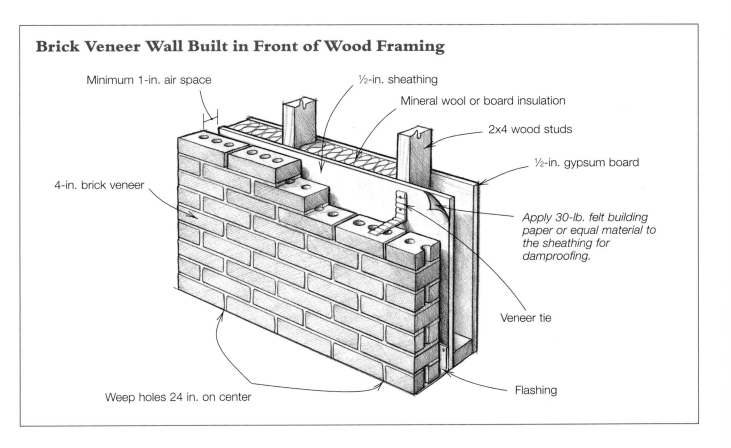

Minimum 1-in. air space

½-in. sheathing

Mineral wool or board insulation

2x4 wood studs

½-in. gypsum board

4-in. brick veneer

Apply 30-lb. felt building paper or equal material to the sheathing for damproofing.

Veneer tie

Weep holes 24 in. on center

Flashing

the wall more energy-efficient. One of the principal advantages for the builder is that since the brick veneer is not a load-bearing wall, the framing of the house can be completed and the brick veneer done later without holding up the other tradesmen.

It is important that weep holes are left at the bottom of the wall in the head joints (maximum spacing 24 in.) to drain any moisture or condensation that may build up inside the wall. Flashing is installed where the weep holes are located. A short length of ⅜-in. clothesline rope is generally placed at the bottom of the head joint to form the weep hole and can be removed later or left in place to act as a wick. The brick veneer facing is tied to the framing by the use of metal wall ties that are spaced every 16 in. vertically and 32 in. horizontally, laid on top of the brick in the mortar joint, and nailed to the wood studs in back. If you are planning to build a brick home or addition, this is the best method to use.

Brick Cavity Walls

A brick cavity wall is usually constructed of a single 4-in. brick facing, a 2-in. minimum to 4-in. maximum air space, and then a 4-in. brick or concrete-block backup wall. The air space in the center can be left open or insulated for greater efficiency. If insulated, a 1-in. separation should be left between the back of the brick and the interior face of the block backing. A 30-lb. felt building paper or equal material should be applied to the sheathing for damproofing. The brick wall is bonded or tied to the masonry backing, using metal Z-ties installed every 16 in. vertically and 32 in. on center horizontally. This type of wall requires step flashing installed across the cavity at the bottom of the wall with weep holes left between the head joints of every third brick in length (24 in.) to drain any moisture or condensation that builds up inside the wall.

Cavity walls are considered to be the most energy-efficient type of brick construction. They do take a little longer to build because of the double-wall design, but

Brick Cavity Wall with Insulation in the Cavity

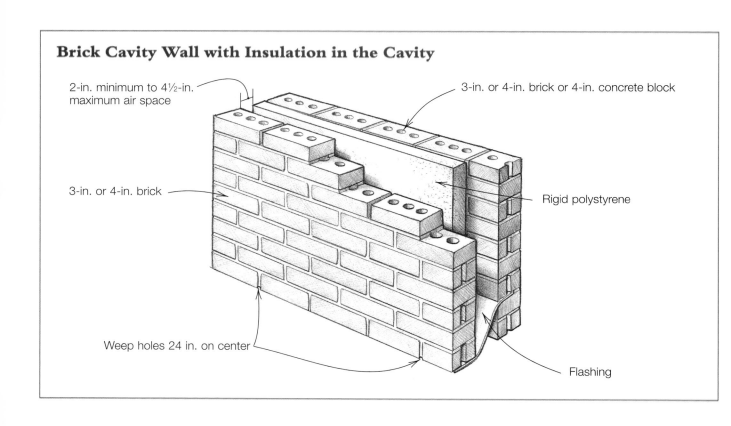

2-in. minimum to 4½-in. maximum air space

3-in. or 4-in. brick or 4-in. concrete block

3-in. or 4-in. brick

Rigid polystyrene

Weep holes 24 in. on center

Flashing

Reinforced Brick Wall with Steel Rods and Grout

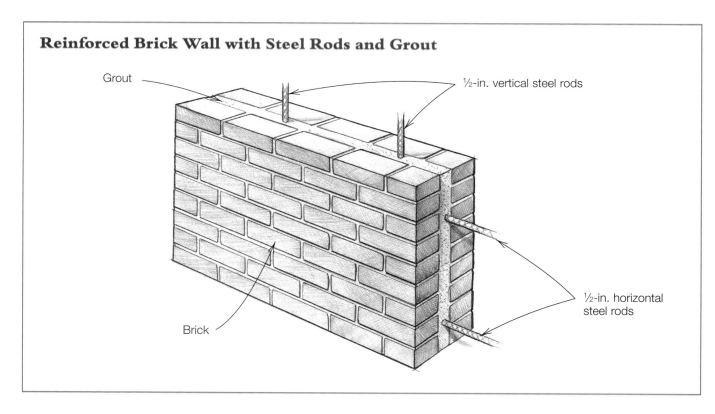

Grout

½-in. vertical steel rods

Brick

½-in. horizontal steel rods

they are well worth the extra time and cost because of the energy savings. With the high cost of heating a home or building, I think that we will see more and more of this type of brick construction in the future.

Reinforced Brick Walls

A reinforced brick wall is the strongest type of masonry wall you can build. A reinforced brick wall is built about the same as a cavity wall with an air space in the center. After the brickwork has cured (that is, the mortar joints have hardened) for at least 24 hours, all openings are sealed off, steel reinforcements are inserted, and the air space is filled with grout. (Grout is similar to concrete but has a different purpose.) Residential projects such as retaining walls and swimming pools are good examples of places where a reinforced wall should be built.

Concrete Block or Brick Retaining Walls

The primary purpose of a masonry retaining wall (shown in the illustration at right and in the photo on p. 56) is to hold back and restrain the earth without cracking or falling out. In my opinion, you cannot overbuild a retaining wall for strength. Most of the time, if a problem develops with a retaining wall pushing out or cracking, the only realistic solution is to tear it down and rebuild it, as you just cannot push it back in place with brute force and expect it to stay put. Once the mortar joints have cracked, the damage has already been done, and makeshift repairs won't do any good. Based on my experiences, a retaining wall for a residential setting, regardless whether it's brick or block, should never be smaller than 8 in. thick—and better yet 12 in. thick for extra strength. It should also be reinforced with steel and grout, with drains behind and through the wall to allow the water that builds up to escape. With all of the hard work it takes to build a retaining wall, this is one project I recommend never cutting corners on! I won't go into a lot more detail here, as Chapter 8 covers how to repair or replace a failed retaining wall.

Serpentine Wall

The *serpentine wall* derives its name from its curving shape, which is in the form of a snake. This unusual repeating curving design provides lateral (horizontal) strength to a brick wall, so that it can be built 4 in. or 8 in. thick without needing any additional support or

Typical Brick and Block Reinforced Retaining Wall

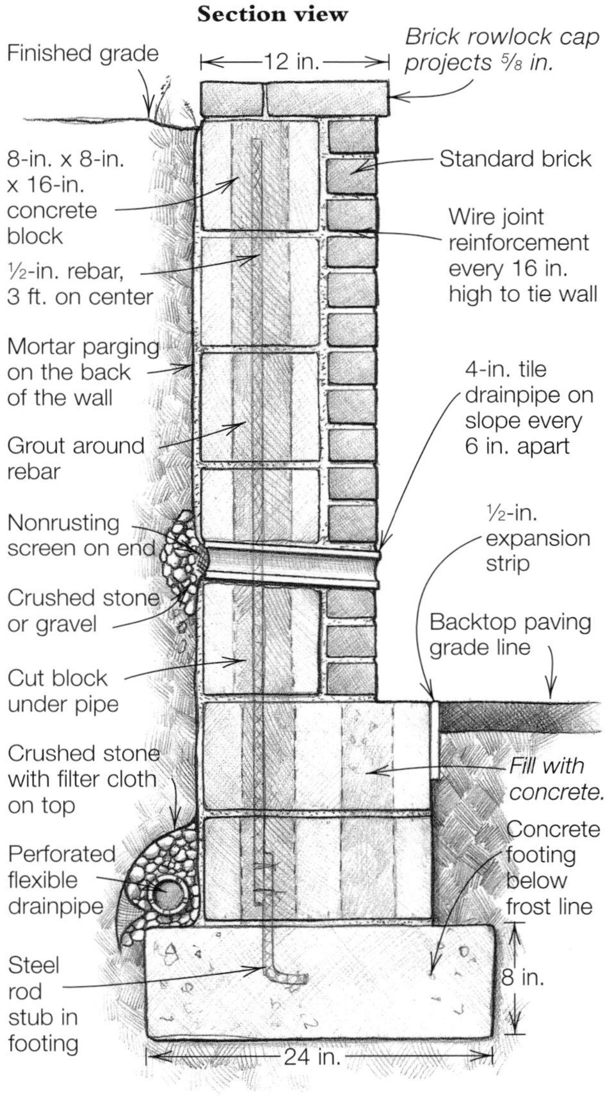

Section view

Finished grade

Brick rowlock cap projects ⅝ in.

12 in.

8-in. x 8-in. x 16-in. concrete block

Standard brick

½-in. rebar, 3 ft. on center

Wire joint reinforcement every 16 in. high to tie wall

Mortar parging on the back of the wall

4-in. tile drainpipe on slope every 6 in. apart

Grout around rebar

½-in. expansion strip

Nonrusting screen on end

Backtop paving grade line

Crushed stone or gravel

Cut block under pipe

Fill with concrete.

Crushed stone with filter cloth on top

Concrete footing below frost line

Perforated flexible drainpipe

8 in.

Steel rod stub in footing

24 in.

Footing is twice as wide as wall.

reinforcement. Granted, you could lay out a garden hose curving it back and forth off a center line and achieve a serpentine wall, but it would not follow an accurate repeating curving pattern.

A brick and block retaining wall lines both sides of a driveway entrance to a street-level garage.

Plan for a Typical Brick Serpentine Wall

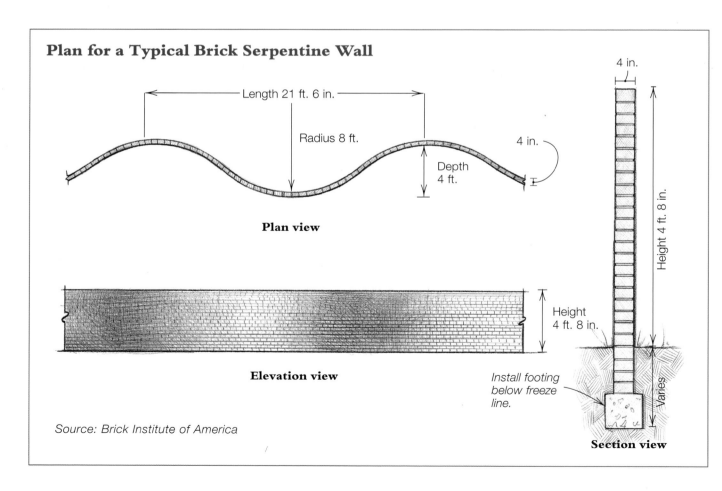

Length 21 ft. 6 in.

Radius 8 ft.

4 in.

Depth 4 ft.

Plan view

4 in.

Height 4 ft. 8 in.

Height 4 ft. 8 in.

Elevation view

Install footing below freeze line.

Varies

Source: Brick Institute of America

Section view

The following general rule from the Brick Institute of America is based on the performance of many successful serpentine walls over the years. The radius of the curvature of a 4-in. brick wall should be no more than twice the height of the wall above finished grade, and the depth of the curvature should be no less than half the height. It is a matter of balance and proportions. It can be built of any masonry material depending on the desired finish and appearance of the wall. The illustration on the facing page provides the details of laying out a serpentine wall.

BRICK AND BLOCK PROJECTS FOR SPECIAL PURPOSES

In addition to laying brick or block to form walls, there are a lot of creative forms of masonry work that can be adapted to fit special needs and purposes. Some are basic projects for learning the fundamentals, and others can be creative, challenging, and just a lot of fun to build. I have included a variety of interesting projects that you may have an occasion to design and build around your home. Although some of the projects don't fit the meaning of traditional walls as we think of them, in one way or another, they make good use of brick and solve existing problems at the same time (see the sidebar on pp. 58-59).

Tree Well Wall

If the grade around a tree has to be filled in or raised a lot, you should consider building a tree well. Minor fills (6 in. or less) don't cause a problem but major ones will. If soil is filled in around a tree to a depth greater than 6 in., the tree will be robbed of its air, water, or minerals, which are necessary for healthy growth. A circular or round brick wall built around the tree provides a well that ensures the tree will receive what it needs (see the top photo on p. 60). A circular brick wall is stronger than a square one, as the stress or pressure acting against it will be applied equally all around.

A circular wall also creates a more pleasing design than a square one and fits into a landscaping scheme better. This is a great project to express your creativity and showcase your masonry skills. It also is an enjoyable project to build. As with any outside masonry wall, you must follow standard building procedures and install a

A brick serpentine wall surrounds a college campus.

concrete footing below the frost line in your area. If you don't, problems of spalling and cracking may result. Also the wall could settle.

You can run into the same problem in reverse—if the grade is cut out too much, the roots may be exposed, which can kill the tree. How much is too much? Generally at least 16 in. of soil should be kept over the roots. The best way to deal with this problem is to build a retaining wall out away from the tree and terrace the grade to maintain the proper amount of soil over the roots.

Brick Piers, Pilasters, and Chases

Brick piers or columns are used in foundations for supporting beams, as supports under porches, and so forth. They also are built to strengthen garden walls, entrances at driveways, and a variety of similar projects. They can be different sizes depending on what you want to support or carry. A cardinal rule I always follow is to lay them out in full brick to eliminate as much cutting as possible and to present an attractive workmanship-like appearance.

Piers are more difficult to build than regular walls because of their small size and because they have eight plumb points (places where a wall is plumbed with a

Typical Brick Pier

level). Care should be taken not to knock piers out of plumb by being too rough.

When a pier is built as part of a wall and projects out or is offset from the face of the wall, it is called a *pilaster*. Pilasters are used a lot in garden and retaining walls to increase lateral support. When they project out on one side of the wall, they are known as *single pilasters* and if they project on both sides, *double pilasters*. The pilaster, like the pier, should be laid out to work full brick and be bonded and interlocked into the brick wall course for more strength.

Brick Pilaster in a Brick Wall

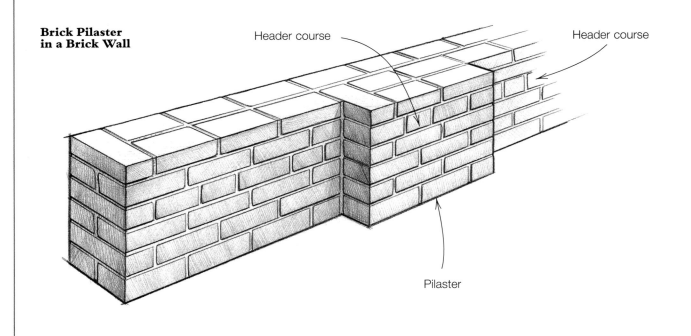

Header course

Header course

Pilaster

A *brick chase* is really the reverse of a pilaster; instead of projecting out, it is recessed in the wall. The brickwork should be constructed and bonded in the brick wall courses the same as the pier and pilaster. Chases are built so that plumbing pipes, downspouts, heating units, and so forth can be set or installed flush with the face of the wall.

**Brick Wall with a
4-Inch-Deep Brick Chase**

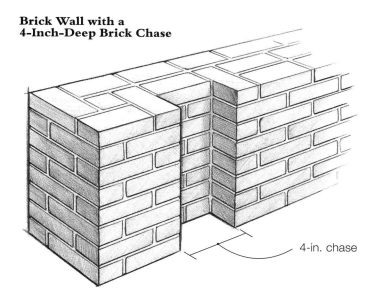

4-in. chase

This brick garden wall has double pilasters.

A brick circular tree well allows a tree to receive the air, water, and soil nutrients it needs to thrive if the grade around it is raised.

Masonry Flower Bed

The grounds around any home can be improved by a brick flower bed. I recommend that the brick wall be at least 8 in. thick to resist weather changes and, of course, be built on a concrete footing below the frost line for your local area. The brick flower bed shown in the photo at right is an excellent example of what you can do with a little imagination and planning. It is built at the intersection of a driveway and concrete steps that go down to the lawn.

The flower bed not only adds some colorful beauty to the area but it also provides a practical solution to turning the corner at the steps. Be sure to build some drains into masonry walls near the bottom, above the finished grade line, to allow excess moisture to drain out. If the last course of solid cap units are laid in alternating flat and on-edge positions, it presents not only a pleasing appearance but also discourages the neighborhood kids from walking on the wall.

A brick flower bed next to the driveway provides an appealing entrance to the home.

Hide that eyesore of a drainpipe with a brick planter built over it.

Brick Planter over Drain

This is a good solution for hiding that ugly drainpipe that goes under a driveway entrance. At the same time, it protects the pipe from being run over by the family vehicle turning into the driveway. You can make the brick wall into something useful like a planter, as shown in the photo above.

Brick Mailbox

If you live in a rural area like I do and constantly have a problem with your mailbox being destroyed by pranksters, try building a brick pier with the mailbox built into it. Make sure that you check with your postal service first and build it to the required height. You also have to set it back off the roadway the correct distance, so as not to interfere with traffic. I recommend that you do not mortar completely around the mailbox inside the pier, but only on the front outside end. This will allow you to replace the mailbox with a minimum of effort if it rusts out.

Building a brick pier around your mailbox protects it against vandalism and is an attractive addition to your property.

In this chapter I have discussed the major bonds and patterns used in brickwork, different types of brick and block walls, and some of the many creative and useful projects that you may want to build around the house. In Chapter 4, I explain the nitty-gritty techniques of laying brick so you can transform all of this information and these ideas into beautiful projects. I strongly recommend that you refer back to the first three chapters as often as needed when you design and build your projects.

Chapter 4
FUNDAMENTALS OF LAYING BRICK

ESTIMATING BRICK,
MORTAR, AND SAND

USING MASON'S
SPACING RULES

TECHNIQUES OF
LAYING BRICK

Watching a professional bricklayer lay brick looks as easy as rolling off a log, but when you try your own hand at it, you soon realize it's not as easy as it looks. The key in learning how to do masonry work is to become familiar with the basic techniques and skills. Then it's a matter of perseverance, and only practice makes perfect.

My brother, Lefty, wanted to build a brick retaining wall in his backyard. He watched me spread the mortar on the wall with the trowel and lay the first couple of courses. After standing back and studying me for a while, he informed me that it looked like a piece of cake and he could handle it. I turned the trowel over to him and was amused to watch the results. The mortar fell off his trowel instead of being spread on the wall, the mortar joints he applied to the end of the brick did not stick, and the brick he laid were sticking out in all directions. He handed the trowel back to me with a grin on his face and the comment, "It sure isn't as easy as it looks!" You'll probably arrive at the same conclusion when you start laying your first brick.

In this chapter I explain each basic step of laying brick and share with you some timesaving tips and techniques such as how to hold a trowel correctly, how to pick up and lay a brick, how to spread mortar on the wall, and how to tool mortar joints. But before you build any brick project, you must have the materials on the job and know how to lay the brickwork out. There-

Keeping a brick wall plumb is important at every step of the project. (Photo © Delmar Publishers.)

fore, it is important to know from the onset how to estimate brick and mortar and how to utilize the mason's scale rules in laying out brick coursing.

ESTIMATING BRICK, MORTAR, AND SAND

As mentioned in Chapter 1, it is important to understand the difference between nominal, specified, and actual dimensions when working with brick. Nominal simply means the brick size plus the mortar joint thickness. The specified dimension is the anticipated manufactured size, and actual dimension is the finished brick size as it comes off the production line. Actual sizes may

vary slightly due to the variations caused by burning and seasoning the brick in the kilns. These size variations are minor and nothing to worry about for most projects.

A good comparison on nominal and actual sizes is found in framing lumber. Even though we say a typical wood stud is a 2x4 (2 in. x 4 in., which is the nominal size), the actual size of the lumber after it has been dressed down is 1½ in. x 3½ in. This is important to know when laying out framed walls, and it's the reason that carpenters lay out from center to center.

In addition to standard modular brick (3½ in. x 2¼ in. x 7½ in.), there are a variety of other sizes manufactured, some of which were discussed briefly in Chapter 1. In case you have an occasion to lay some of these brick, you could use the chart on p. 64 to estimate the number of brick and amount of mortar needed for these different sizes.

The chart on p. 65 and the drawing on p. 66 of modular and nonmodular brick show the sizes and appearances of these brick. It is unlikely that you will ever use all of these different brick sizes shown, but I've included them here as a handy reference guide if you ever do need to use them.

Brick

There are two ways to estimate brick. One is *rule-of-thumb estimating*, which provides a rough estimate and was developed from years of practical experience on the job and then checked when the job was complete to determine if it worked out. The second way is estimating by using specific established tables and slide rules and is normally used for estimating large jobs. Rule-of-thumb estimating works great for projects or jobs up to the size of an average house. It is not intended to be a perfect mathematical solution, but it will more than suffice for the average job. One of the benefits of rule-of-thumb estimating is that it allows for a little waste or overrun.

Rule-of-thumb estimating starts with determining the square footage of the wall or project. This is known as the *square-foot-wall-area method*. If the total square footage works out to a fractional number, it is best to

Modular and Nonmodular Solid Brick Units with Mortar for Single Wythe Walls in Running Bond

	Brick Size (in inches)			Brick per 100 sq. ft.		Cubic Feet of Mortar per 1,000 Brick	
Modular (nominal)	t	h	l			⅜-in. joint	½-in. joint
	4	2⅔	8	675		8.1	10.3
	4	3⅕	8	563		8.6	10.9
	4	4	8	450		9.2	11.7
	4	5⅓	8	338		10.2	12.9
	4	2	12	600		10.8	13.7
	4	2⅔	12	450		11.3	14.4
	4	3⅕	12	375		11.7	14.9
	4	4	12	300		12.3	15.7
	4	5⅓	12	225		13.4	17.1
	6	2⅔	12	450		17.5	22.6
	6	3⅕	12	375		18.1	23.4
	6	4	12	300		19.1	24.7
	8	4	12	300		25.9	33.6
Nonmodular (actual)	t	h	l	⅜-in. joint	½-in. joint	⅜-in. joint	½-in. joint
	3	2⅝	8⅝	532	505	7.6	9.7
	3	2⅝	9⅝	481	457	8.2	11.1
	3	2¾	9¾	457	433	8.4	11.3
	3	2¼	10	529	500	8.2	11.1
	3¾	2¼	8	655	616	8.8	11.7
	3¾	2¾	8	551	522	9.1	12.2

Notes:
A nominal size or dimension is the actual dimension of the unit plus the joint thickness.
t = thickness; h = height; l = length
Source: Brick Institute of America

round it off to the next highest whole number. Be sure to deduct the total area of any windows, doors, or openings from the square foot figure. When estimating brick, it's always better to have a few extra to cover the chipped ones. Mortar head and bed joints are always included in the square footage, so there is no need to allow any extra for them. The following example is based on using standard brick, which measures approximately 3⅝ in. x 2¼ in. x 7⅝ in. and is laid up with ⅜-in. mortar head and bed joints.

Let's assume that you're going to build a brick wall 8 ft. high and 20 ft. long. To simplify things, it has no open-

Modular and Nonmodular Brick Sizes

Unit Designation	Nominal Dimensions (in inches)			Joint Thickness** (in inches)	Specified Dimensions*** (in inches)			Vertical Coursing
	w	h	l		w	h	l	
Modular Brick Sizes*								
Modular	4	2 2/3	8	3/8 1/2	3 5/8 3 1/2	2 1/4 2 1/4	7 5/8 7 1/2	3C=8 in.
Engineer Modular	4	3 1/5	8	3/8 1/2	3 5/8 3 1/2	2 3/4 2 13/16	7 5/8 7 1/2	5C=16 in.
Closure Modular	4	4	8	3/8 1/2	3 5/8 3 1/2	3 5/8 3 1/2	7 5/8 7 1/2	1C=4 in.
Roman	4	2	12	3/8 1/2	3 5/8 3 1/2	1 5/8 1 1/2	11 5/8 1 1/2	2C=4 in.
Norman	4	2 2/3	12	3/8 1/2	3 5/8 3 1/2	2 1/4 2 1/4	11 5/8 11 1/2	3C=8 in.
Engineer Modular	4	3 1/5	12	3/8 1/2	3 5/8 3 1/2	2 3/4 2 13/16	11 5/8 11 1/2	5C=16 in.
Utility	4	4	12	3/8 1/2	3 5/8 3 1/2	3 5/8 3 1/2	11 5/8 11 1/2	1C=4 in.
Nonmodular Brick Sizes								
Standard				3/8 1/2	3 5/8 3 1/2	2 1/4 2 1/4	8 8	3C=8 in.
Engineer Standard				3/8 1/2	3 5/8 3 1/2	2 3/4 2 13/16	8 8	5C=16 in.
Closure Standard				3/8 1/2	3 5/8 3 1/2	3 5/8 3 1/2	8 8	1C=4 in.
King				3/8	3 3	2 3/4 2 5/8	9 5/8 9 5/8	5C=16 in.
Queen				3/8	3	2 3/4	8	5C=16 in.

Notes:
* *1 in. = 25.4 mm; 1 ft. = 0.3 m*
** *Common joint sizes used with length and width dimensions. Joint thicknesses of bed joints vary based on vertical coursing and specified unit height.*
*** *Specified dimensions may vary within this range from manufacturer to manufacturer.*
w = width; h = height; l = length
Source: Brick Institute of America

Sizes and Appearances of Modular and Nonmodular Brick

Modular brick sizes (nominal dimensions)

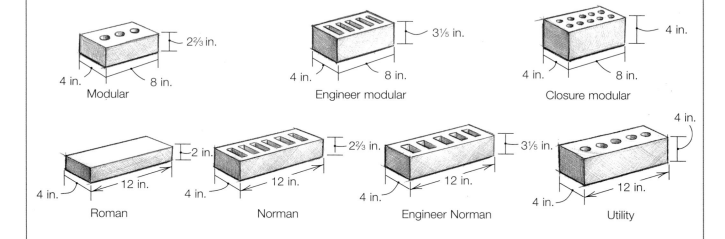

Modular — 4 in., 8 in., $2\frac{2}{3}$ in.

Engineer modular — 4 in., 8 in., $3\frac{1}{5}$ in.

Closure modular — 4 in., 8 in., 4 in.

Roman — 4 in., 12 in., 2 in.

Norman — 4 in., 12 in., $2\frac{2}{3}$ in.

Engineer Norman — 4 in., 12 in., $3\frac{1}{5}$ in.

Utility — 4 in., 12 in., 4 in.

Nonmodular brick sizes (specified dimensions)

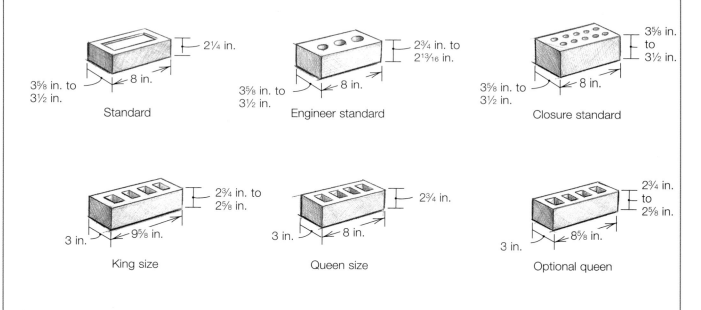

Standard — $3\frac{5}{8}$ in. to $3\frac{1}{2}$ in., 8 in., $2\frac{1}{4}$ in.

Engineer standard — $3\frac{5}{8}$ in. to $3\frac{1}{2}$ in., 8 in., $2\frac{3}{4}$ in. to $2\frac{13}{16}$ in.

Closure standard — $3\frac{5}{8}$ in. to $3\frac{1}{2}$ in., 8 in., $3\frac{5}{8}$ in. to $3\frac{1}{2}$ in.

King size — 3 in., $9\frac{5}{8}$ in., $2\frac{3}{4}$ in. to $2\frac{5}{8}$ in.

Queen size — 3 in., 8 in., $2\frac{3}{4}$ in.

Optional queen — 3 in., $8\frac{5}{8}$ in., $2\frac{3}{4}$ in. to $2\frac{5}{8}$ in.

Source: Brick Institute of America

ings. First, determine the square footage by multiplying the height by the length (8 ft. x 20 ft. = 160 sq. ft.). To find out how many brick are needed, multiply 160 by 7 (there are approximately 7 standard brick in every square foot of wall). This allows for a small waste factor because technically there are exactly 6.75 standard brick in 1 sq. ft. of a single *wythe* (each continuous vertical section of a masonry wall or one unit in thickness) brick wall. The total number of brick needed for this example is 1,120. Remember, if building a double brick wall, you will have to double this amount. That's all there is to it!

Mortar

Estimating the amount of mortar needed for laying brick is relatively simple. There are basically two kinds of mortar you can use, which were discussed briefly in Chapter 1. The one most frequently used is masonry cement mortar. All you need to add is sand and water. Estimate that one bag of masonry cement mortar will lay 125 brick when mixed with 18 shovelfuls of sand. When mixing smaller amounts, use a ratio of 1 part masonry cement to 3 parts sand. Referring to the number of brick in the example above, it would take 8.96, or 9 bags to lay the 1,120 brick.

If you are doing a small job that will only require several bags, preblended mortar that already has the sand added is a good choice. One 40-pound bag will lay approximately 50 brick. If you decide to use a portland-cement/lime mortar mix, estimate that a full batch—one bag of Type 1 portland cement to one bag of mason's hydrated lime to 42 shovelfuls of sand—will lay approximately 300 brick. If you want to mix a smaller amount, the ratio would be 1 part portland cement to 1 part hydrated lime to 6 parts sand. When working by yourself or on a small project, it is wise to mix only a half batch at a time, as the mortar will have the tendency to dry out before being used. Mortar should be discarded 2½ hours after the initial mixing. After that, the strength is greatly reduced.

Sand

Sand is the least costly of all mortar ingredients. To estimate the amount of washed building sand needed for brick mortar, figure 1 ton of sand for every eight bags of masonry cement mortar. This amount will lay about 1,000 brick. You can buy sand in lesser amounts, by the pound or in 80-pound to 100-pound bags for smaller projects; however, it is more expensive this way. The previous example of 1,120 brick would require about 1¼ tons of sand.

USING MASON'S SPACING RULES

As mentioned in Chapter 2, masons use two principal types of folding spacing rules to divide and evenly space the height of individual courses of brick or masonry units—the brick mason's spacing rule and the modular spacing rule. These scales always include the mortar bed joint thickness. Before building any brickwork, you need to understand how to use these rules to work out the individual courses. I strongly recommend that you purchase one of each, as you will run into modular and nonmodular brick on many occasions. Both are shown in the photo below.

Brick Mason's Spacing Rule

This spacing rule is a standard 72 in. long and was developed before the modular system of building was created. (The term *modular* simply refers to materials such as brick, concrete block, tiles, or other building materi-

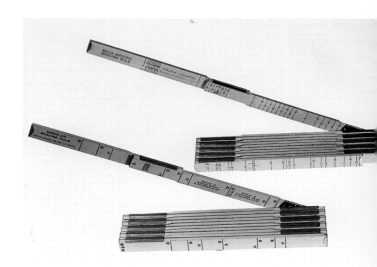

Standard spacing rules—the mason's spacing rule (*top*) and the modular spacing rule (*bottom*)—are 72 in. long.

als that are manufactured to fit into a 4-in. grid or multiples of 4 in. Architects and engineers developed this grid to minimize cutting or sizing. Most building materials made today are modular.) It was designed specifically to allow for dividing mortar joints evenly in a wall that does not work out to fit the modular grid. This problem occurs a lot in old brickwork or when additions or renovations are made to older buildings.

Here's a basic example to illustrate how this works. Suppose you are bricking around a window opening in a wall, and the overall height of the window frame is 46¾ in. Since the modular rule only works evenly in modules of 4 in., it cannot be used to mark off brick courses that are exactly equal on this frame. The solution is to use the spacing rule and adjust the difference. If you examine the standard side of the rule and then turn it over, you'll notice that the 46¾-in. figure lines up exactly with No. 6 on the scale side. You will also see that on scale No. 6 at this point, the number of courses (17 in this case) are marked in red, which is very helpful when estimating the number of brick needed. (*Caution:* No. 6 on the spacing rule does not mean the same thing as No. 6 on the modular rule, which I will explain later.) Using this scale, you will find that all of the mortar bed joints need to be slightly thicker than the normal ⅜ in. to make this work out in even courses of brick. This small sequential increase in the height of the mortar joints, if done the same on each course, is seldom noticeable to the average eye.

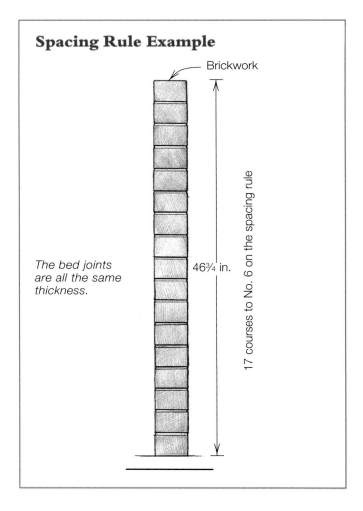

Spacing Rule Example

Brickwork

The bed joints are all the same thickness.

46¾ in.

17 courses to No. 6 on the spacing rule

Keep in mind that the spacing rule was designed to be used only for standard brick heights. Only the mortar bed joint thickness changes, not the brick itself. This same application can be utilized when laying out a brick windowsill. All you need to do is turn the rule in a horizontal instead of vertical position and mark off the individual brick on the wood sill of the window frame to use as a guide when installing the sill.

The drawing on the facing page further illustrates how mortar bed joints can be increased or decreased evenly by following a scale and be made to work out to any dimension. The larger the number is on the spacing rule, the thicker the mortar joint will be.

Mortar bed joints are not always made thicker. They can also be made thinner, which is sometimes necessary to make the brick courses work out evenly. There is, of course, a limit on how thin you can make a bed joint without compromising its strength. My recommendation is not to drop below No. 4 on the scale for thinner joints and above No. 6 on the scale for thicker joints. As a general rule, mortar joints should be ⅜ in. minimum to ¾ in. maximum for exterior wall construction.

An important feature of the mason's spacing rule is that the total number of brick courses from the bottom to the top of the rule is indicated on the scale side. This is

Application of Brick Mason's Spacing Rule

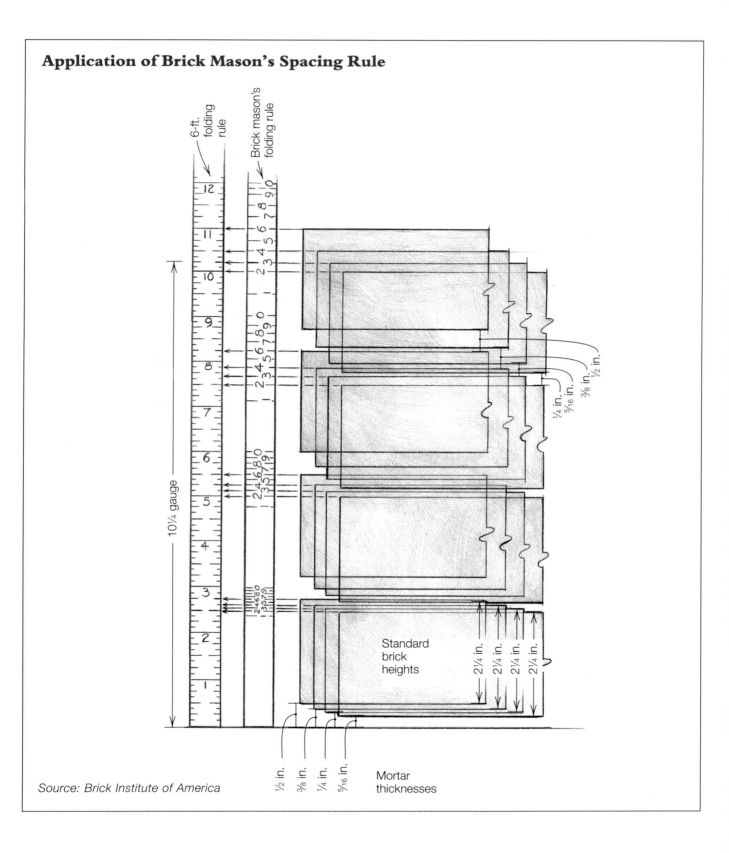

6-ft. folding rule

Brick mason's folding rule

10¼ gauge

Standard brick heights

2¼ in. 2¼ in. 2¼ in. 2¼ in.

¼ in. 5/16 in. 3/8 in. ½ in.

½ in. 3/8 in. ¼ in. 5/16 in.

Mortar thicknesses

Source: Brick Institute of America

very helpful when you want to estimate the number of courses of brick in a wall or project. The height of a wall can be checked, while at the same time you can quickly determine the number of courses needed. For example, a brick project that is 71½ in. high will require 26 courses of brick laid to No. 6 on the spacing rule.

Modular Spacing Rule

As with the brick mason's spacing rule, the modular spacing rule has the standard 72 in. on one side and a scale on the opposite side. (The modular rule is shown in the photo on p. 67.) This scale can only be used with brick that fit into the modular grid of 4 in. The courses of standard brick work out to No. 6 on the modular scale because there are six courses of brick to every 16 in. in height. The accepted height in the United States for modular masonry units or courses to level off is 16 in. This is because every 16 in. vertically requires a

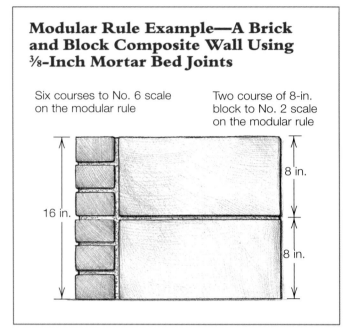

Modular Rule Example—A Brick and Block Composite Wall Using ⅜-Inch Mortar Bed Joints

Six courses to No. 6 scale on the modular rule

Two course of 8-in. block to No. 2 scale on the modular rule

16 in.

8 in.

8 in.

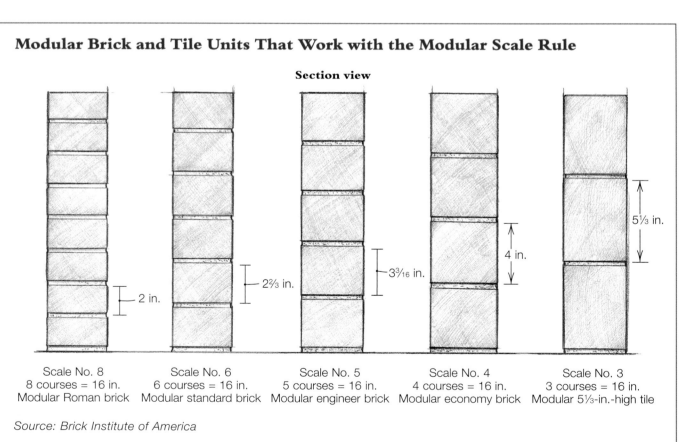

Modular Brick and Tile Units That Work with the Modular Scale Rule

Section view

2 in.

2⅔ in.

3³⁄₁₆ in.

4 in.

5⅓ in.

Scale No. 8	Scale No. 6	Scale No. 5	Scale No. 4	Scale No. 3
8 courses = 16 in.	6 courses = 16 in.	5 courses = 16 in.	4 courses = 16 in.	3 courses = 16 in.
Modular Roman brick	Modular standard brick	Modular engineer brick	Modular economy brick	Modular 5⅓-in.-high tile

Source: Brick Institute of America

metal tie to bond the wall together properly. An example of this is six courses of standard brick backed up by two courses of concrete block. Since a concrete block is 8 in. in height, including the mortar bed joint, and it takes two courses to equal 16 in. vertically, a course of block would be represented as No. 2 on the modular scale rule.

Each multiple of 16 in. is usually marked in red on the standard side of the modular scale rule to remind the user that different kinds of masonry materials will level off or coordinate at this height. This follows the same principle as the carpenter's steel tape, which has a diamond mark every 16 in. in length to indicate the modular layout of lumber such as wood 2x4s for studs or rafters. To make the scales clear to the user, the types of masonry units that apply to the modular scales are printed on the first 24-in. section of the scale side of the modular scale rule.

The bottom drawing on the facing page illustrates a section view of modular brick and tile units that work with the modular scale rule. You can see how these units fit into the modular scale.

TECHNIQUES OF LAYING BRICK

Before starting any brick project, you need a plan or drawing to help avoid mistakes. You should be able to visualize in your mind what the finished project will look like before laying that first brick. (See the sidebar on pp. 72-73.) Spending a little extra time in the beginning thinking things through will pay big dividends in cutting down on potential problems and having to tear up and relay the brickwork. There is an old saying in carpentry work that you should always "measure twice and cut once." In brickwork, it is better to make the first mistakes on paper than on the wall.

Leads and Corners

Any brick project requires building an end or corner of some type before building the wall. There are three basic types of ends or corners (called *leads* in the trade) that can be built to serve as guides for a project. Incidentally, in the masonry trade, each layer of brick or block is called a *course,* instead of layer or row.

Rack-back lead The simplest lead is known as a *rack-back lead,* which is not technically a corner because it does not return on an angle but serves as a guide to attach a line to when laying the brick wall. The rack-back lead consists of a number of brick courses that are built plumb and level to a specific height and racked back one-half brick on each end of every course until it cannot be built any higher. The first course should not be laid out longer than six standard brick (48 in.). There is a logical reason for this. If the lead is the same length as the 4-ft. level, this helps to prevent building the lead out of line. It is poor practice to

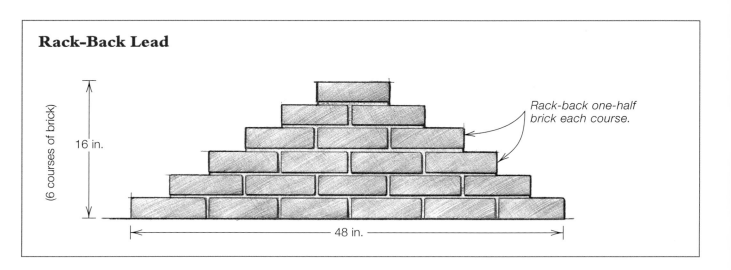

Rack-Back Lead

(6 courses of brick)

16 in.

Rack-back one-half brick each course.

48 in.

Laying Out a Course Rod

By now you should have a basic understanding of how the mason's scale rules are used. Let's apply this information and make a *course rod* (or *story pole*) for a project. Using the example of a brick garden-type storage building in the drawing in this sidebar, let's lay out a typical course rod. Since the vertical wall measurements are laid out in multiples of 4 in., the modular rule will be used.

1. Select a reasonably straight wood rod free of any large or loose knots, such as a 1x3 furring strip. If you don't have one lying around, you can buy one at any home center.

Referring to the example, notice that the overall height of the wall is 96 in. Using the modular scale rule, you'll find that this works out to perfect courses on scale No. 6 for a total of 36 courses of brick high. Square a line across on the rod at the 96-in. height, and cut the rod to the line with a handsaw. Using a sharp pencil, mark the top of this line with a crow's foot (arrow), and print "top" so the pole does not get turned upside down accidentally when using it. Holding the rule even with the bottom of the rod, measure up and mark the top of the window and door. Square this mark across with a pencil and print "WH" (window height) and "DH" (door height). As you can see on the example, they are both at the same height for

simplicity. This measurement is 80 in. and works out to 30 courses of brick.

2. Measure up from the bottom of the rod, using the square and pencil. Mark the top of the brick windowsill, which is 32 in., or 12 courses of brick high. Since this brick sill is laid in a header position (flat with the end showing), it will work out to regular course height. (Brick positions were discussed in Chapter 3.) Print "sill" at the inside top of this line, and draw an X across it to remind you that this is where the sill for the window goes.

3. Mark the individual courses of brick to No. 6 on the modular rule from the bottom to the top of the rod. Using the square, extend each of these lines across the rod.

4. Starting at the bottom of the rod again, number each course of brick to the top. Be sure to recheck all measurements for accuracy when you are finished. In this case, it will work out to 36 courses high.

After the course rod is finished, it is always a good idea to make the marks permanent by sawing them in lightly to a depth of ⅛ in. with a handsaw. This prevents any problems with the marks fading.

Shorter versions of a course rod can be made to fit smaller projects and are called *sill rods*. A

course rod saves a lot of time constantly unfolding and folding a rule. (Be especially careful when you unfold or fold a rule, as the joints have nasty pinch points that will raise a king-size blood blister.) The course rod also is more accurate than using a folding rule, as the rule may bend when held vertically. When building any project over 6 ft. high, which is the full length of an extended mason's rule, I

Brick Garden Storage Shed Elevation Plan in Modular Scale

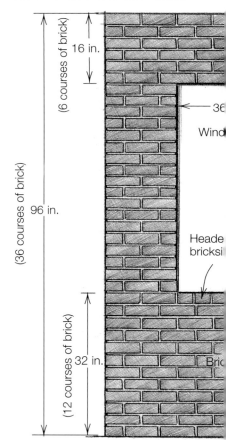

16 in. (6 courses of brick)

96 in. (36 courses of brick)

32 in. (12 courses of brick)

36

Wind

Heade bricksi

Bri

strongly recommend making a course rod, as there is a heck of a lot less chance for error.

Keep in mind that when using a course rod, the bottom of the rod must be set at the same level base point on all corners. You can establish these level points in different ways. For example, you can use a long straightedge board with a level laid on top of it, a transparent garden hose with water in it (water will always lie level), or for a larger project, a builder's level. Once established, I usually mark and drive a nail in the mortar joint at these level base points and set the course rod on the nail when checking course heights. These level points are commonly known as *benchmarks* in the trade.

After putting all of this effort into making a course rod, be sure to take good care of it during the building of your project by keeping it dry and not laying anything on top of it. This will prevent any bowing or warping. Regardless of the size of brick or concrete block you're laying out, the basic procedure for making a course rod is the same. The only thing that changes is the size and joint thickness of the materials.

Course rod matches heights on shed.

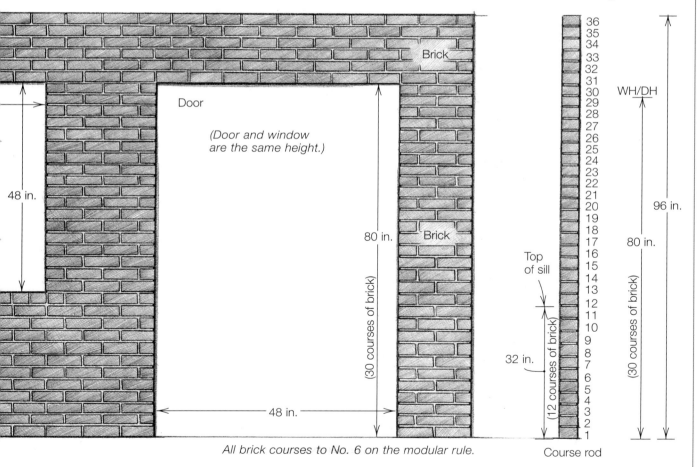

All brick courses to No. 6 on the modular rule.

Straight Lead

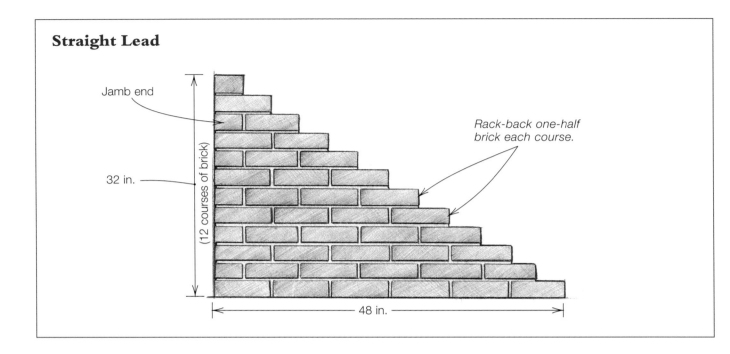

Jamb end

32 in.

(12 courses of brick)

Rack-back one-half
brick each course.

48 in.

build long rack-back leads because the brick can be laid faster and more efficiently to a line (if the lead is no longer than 48 in.).

Straight lead The second type is known as a *straight lead*. It is simply a rack-back lead that is built plumb on one end. This is a little tougher to build than the rack-back lead because one end as well as the face of the lead must be plumb. There is a rule that you should always follow when building any lead or corner—plumb a single brick wall only on the finished side because of slight differences in brick widths. An example of when you would build a straight lead is at the end of a wall or next to an opening such as a door or window jamb. This lead can be built higher than a rack-back lead because you only rack back half a brick on one end. For example, a 6-brick layout on the first course would result in a completed straight lead of 12 courses of brick after racking back. (See the drawing above.)

Right-angle corner The third type is a standard 90-degree right-angle corner. In reality, it consists of two straight leads joined together to form a corner. Since

the basic skills required to build leads and corners are the same, I have selected the right-angle corner as a project to describe in detail.

Building a Right-Angle Corner

The first step is to brush off the base and mark off the layout lines using a 2-ft. framing square (see the left photo on the facing page). It is a good idea to extend these two lines out a couple of extra feet with a straight-edge and snap a chalk line on the base.

Dry-bonding the first course Keep in mind that the total number of layout brick on the first course will equal the number of courses high in the completed corner. Therefore, a corner nine courses high will require five brick laid out in one direction and four in the opposite direction.

Dry-bond (lay out without mortar) the first course of brick along the layout line. I use the tip of my little finger held flat as a spacer between each brick. In most cases this will be approximately ⅜ in. This does not have to be an exact measurement. Lay the straightest or best side of the brick to the layout line, as this will be the

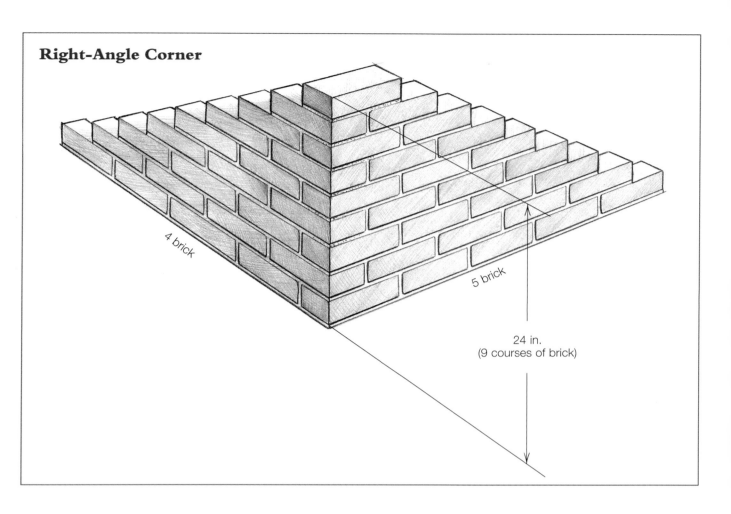

Right-Angle Corner

4 brick

5 brick

24 in.
(9 courses of brick)

The first step in building a right-angle corner is to square the first course on the base, being done here with a square and chalk line.

To space out the block for dry-bonding, use your little finger as a spacer.

Cutting and Spreading Mortar

Holding a trowel with the end of your thumb on top and over the front of the handle a bit prevents you from burying your thumb in mortar. (Photo © Delmar Publishers.)

I waited until now to discuss cutting and spreading mortar so I didn't put the cart before the horse. First of all, there is a right way and a wrong way to hold a brick trowel. I recommend holding the trowel in your hand with the end of your thumb on top and slightly out over the front of the handle with your fingers gripping the rest of the handle. This prevents you from burying your thumb in the mortar on the board, which may cause the skin to crack or split open. It does not make any difference if you are right- or left-handed.

There are several ways of cutting and picking up mortar with the trowel from the board. The simplest and easiest to master is called *cupping*. I have broken it down into four basic movements.

1. To cut a trowel of mortar free from the main pile on the board, slice down with the trowel blade and pull free an amount that would approximately fill the trowel. With a rolling motion, pull this smaller pile to the edge of the board.

2. Shape the mortar into the form of a cigar or roughly the shape and length of the trowel blade using the trowel. This will ensure that the trowel is filled with mortar and will make it easier to spread uniformly.

3. Slide the trowel blade sideways under the mortar. This will help the mortar stick to the trowel blade.

4. Pick up the trowel, snapping your wrist downward lightly to set the mortar on the trowel blade. Don't overdo it or the mortar may fall off. You should now have a properly filled trowel of mortar ready to spread.

Cutting the mortar from the main pile. (Photo © Delmar Publishers.)

Shaping the mortar to fit the trowel blade.

Sliding the trowel blade under the mortar.

A trowel properly loaded with mortar.

finished face of the wall. Mark where the end of each brick goes on the outside edge of the layout line with a pencil. Then push these brick back off the line in preparation for spreading mortar on the base.

Laying the first course in mortar Pick up a trowel of mortar and with a relaxed swing of the arm, spread the mortar along the line as evenly as you can. Keep the point of the trowel about even with the middle of where the brick will be laid and the blade turned over to allow the mortar to release smoothly. I would compare this to the backward motion of sowing a handful of grass seed on the lawn. At the completion of the spread, fill in any bare spots with mortar. If you have properly cut mortar from the board and have the trowel loaded the way I described in the sidebar on pp. 76-77, it should spread out reasonably well. Try not to cover up the line. It may take you several tries to get the hang of it.

Next, furrow (make an indentation) the center of the mortar bed with the point of the trowel. Do this by holding the trowel at approximately a 45-degree angle and moving the trowel with a slight hopping motion of the wrist. Work it through the length of the mortar bed. Don't punch so hard that the trowel goes all of the way through to the base, which could leave a void and cause the joint to leak later on. When finished, pull any mortar that is covering the line back with the trowel. Furrowing is done to even out the mortar bed and to make it easier to press the brick down when laying them.

Lay the corner brick to No. 6 on the modular rule. Then lay the brick on the opposite end of the course level with the corner brick. Lay the brick in between in mortar just a shade high.

Now level the course of brick from the corner point out. Tap down any high brick or relay any low brick by taking the low brick up and adding mortar. (It is best to

Spread mortar along the line with the trowel.

Furrow the mortar bed to even out the mortar and settle the brick easier.

Use the 4-ft. level to plumb the first course of brick.

Plumb the face side of the brick, adjusting any brick that are positioned incorrectly.

Straightedge the course along the top edge.

re-lay any brick that are moved or disrupted once initially placed.) Be sure to lay the brick with the face or best side out.

Plumb the face of both the corner and the end brick with the level. Then align the course by holding the edge of the level horizontally against the top outside edge of the brick. Tap back any brick that stick out and adjust any brick that are slack. This is known as *straightedging*. It is important to always line up the brick at the top edge, as many brick have a slight lip there. This helps to keep the wall in line.

Applying the mortar head joint As you lay brick, a mortar head joint must be applied to the end of the brick to receive the next one laid against it. There are two methods of doing this. You can either swipe the mortar on the end of the brick while holding it in your hand, or you can hook the mortar head joint with the trowel on the end of the brick after it is laid in the wall. At this point I recommend swiping the mortar joint on the end of the brick while holding it in your hand. Later, when laying brick to a line, the other method is more productive.

This is done in two basic movements. Holding the brick at about a 45-degree angle, pick up a small amount of mortar on the front end of the trowel, setting it on the blade with a little snap of the wrist (see the top left photo on p. 80). Then holding the brick in the opposite hand on approximately a 45-degree angle and with a downward slicing motion of the trowel, swipe the mortar onto the end of the brick. The impact of the mortar striking the end of the brick will cause it to stick. (Notice in the top right photo on p. 80 that the hand is holding the brick in the center to prevent being cut.) This is not difficult to master—it just takes a little practice.

Seal off the bottom of the bed joint where it meets the base by smoothing it out with the point of the trowel held flat against the bed joint (see the bottom photo on p. 80). This prevents any water from leaking through.

Repeat the same procedure on the opposite side of the corner. I have found that it helps to lay the last brick on the end and then fill in between on the first course in the event that the base is not exactly level.

Pick up mortar on the front edge of the trowel, preparing to apply it to the brick, for a head joint.

Swipe mortar on the end of a brick.

Seal the bed joint with the point of the trowel.

Continue to lay succeeding courses of brick, checking the height of each course to scale No. 6 on the modular rule. After the first course of brick is laid level, insert the trowel blade in the mortar bed joint on top of the first brick course and rest the rule on top of the blade to check the height. I recommend doing this because sometimes the base under the first course may be low and may have required a big mortar bed joint to start off.

Tailing the end of the corner After the last brick have been laid and the corner completed, it is a good idea to *tail* (line up on an angle) the ends of the corner. This is done by holding the level on a diagonal against the ends of the brick on the first and last courses. If any of the brick are out of line, adjust them to the edge of the level. Don't adjust the bottom brick or the corner will be out of its original layout line. Repoint any hairline cracks that occur with fresh mortar.

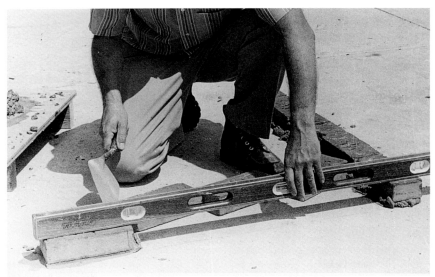

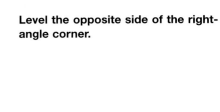

Level the opposite side of the right-angle corner.

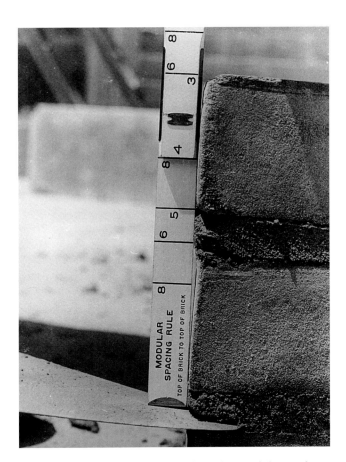

Check the course height to No. 6 on the modular scale as you lay succeeding courses of brick.

Tail the ends of the corner with the level, tapping the brick into line.

Good Practices to Follow when Laying Brick

• When laying brick in the mortar bed, face the wall and pick up the brick in your hand, with the thumb and first and second fingers resting on the front and back of the brick. Press the brick down gently in the mortar bed joint, lining it up as close as possible to the face of the wall below. Pressing the brick down with equal pressure from all three fingers helps position the brick better. Don't drop the brick into the bed joint, but press it down to where you think it is approximately level.

Move your hand back and lay it flat, so that half extends over the brick previously laid and half over the brick being laid. If you are not wearing gloves, you can feel in the palm of your hand when the brick is close to being level. At the same time, holding the trowel on a slight angle to prevent smearing the face, cut off the excess mortar being squeezed out and use it to apply the next mortar head joint on the brick.

After each course is laid, continue to set the level in the middle of the course and tap down any high brick with the edge of the trowel. Low brick must be taken up and relaid with additional mortar until level. Be careful not to tap along the front edge of the brick as it may chip the face! Never tap or beat on top of the level as you may damage the vials where the bubble is located.

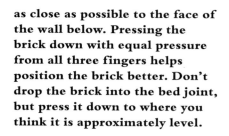

Lay the brick in position using two fingers and thumb.

• A skill that will greatly speed up your bricklaying is the ability to sight vertically down the wall and lay the brick as close to plumb as your eye tells you. This reduces adjusting the brick in and out when you plumb the wall with the level.

Try this: Stand directly over the brick as you lay them and sight down the corner or wall. Line the brick up with the ones below as close to plumb as you can as you cut the mortar bed joint off with the trowel. Believe it or not, as little as ¼ in. out or in can be seen readily and an adjustment can be made right then. This technique alone, once mastered, will tremendously cut down the amount of time you spend plumbing a project.

• When the weather is hot and humid, mortar tends to dry out prematurely from evaporation or from lying too long on a dry wood mortarboard. You can restore it to workable condition by adding a little water and remixing it on the board with the trowel. This is called *tempering* in the trade. Do not add so much water that the mortar becomes thin or runny. A good tip to follow is don't add any more water than was lost by

Press the brick in position and cut excess mortar. (Photo © Delmar Publishers.)

Sighting down the corner can readily show brick even ¼ in. out of line. (Photo © Delmar Publishers.)

evaporation. I also recommend not tempering mortar more than twice or its bond strength will be weakened. If you don't mix more mortar that you can use in one hour, you will always have full-strength mortar with a minimum of tempering.

• Brick will lay a lot easier in hot weather if they are dampened a little with water. It is easy to tell when this is necessary as the brick will not press down readily in the mortar joints when laid. This is especially true with a sand-finish brick. Don't soak the brick; just sprinkle some water on them using a fine spray from a hose or a water bucket and a brush. You will be pleasantly surprised at how much easier brick will settle in the mortar joint. However, I recommend never wetting brick in cold or freezing weather, as the brick or mortar could crack if freezing takes place before the mortar dries out.

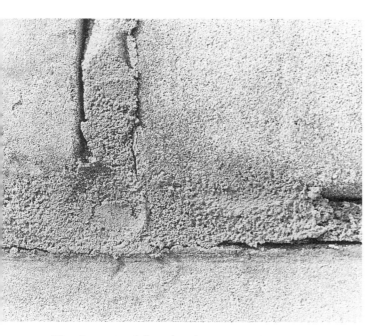

Check mortar joints for the proper time for tooling by pressing your thumb into the mortar joint.

Pick up mortar on a convex jointing tool by pressing downward and rolling your wrist.

Tooling or striking the mortar joints In Chapter 2, I mentioned a number of jointing or striking tools. Mortar joints are *tooled* (finished) to seal and beautify them. Since drying conditions of the joints vary, before tooling I recommend testing the mortar joint periodically by pressing your thumb in the mortar. If your thumb leaves an impression, then the joint is dry enough and should be tooled immediately. Tooling joints when they are too wet can cause excessive smearing or sagging. On the other hand, if you wait too long and the mortar is past what is called its *initial set,* the metal jointing tool can cause black marks in the joint. This is commonly called *burning the joints* and is unsightly.

Pick up a small amount of mortar on the front part of the trowel and with a slight snap of the wrist, set it on the blade trowel so it will stick. I typically use a convex jointer that forms a half-round impression in the mortar joint (see the bottom photo at left). Hold the tip of the trowel up slightly, so the mortar will slide to the rear. Pick up the mortar from the trowel on the edge of the jointing tool by pressing downward and at the same time rolling your wrist up. If done correctly, this pressing down will cause the mortar to stick to the jointer.

I recommend always tooling the head joints before the horizontal bed joints. The reason for this is that tooling the head joints leaves a slight overrun or impression in the bed joint. You can remove these imperfections by running the jointing tool through the bed joint last. The finished bed joints should be in a straight line without any imperfections or marks in them.

To tool the head joints, hold the jointer vertically and pull it through the joints with the curved edge in the center of the joint. Fill any holes or voids in the joints as you do this.

To tool the bed joints, hold the jointer horizontally, and pressing firmly, pull it through the length of the joint smoothly so that the mortar fills out evenly to all of the edges, leaving no gaps anywhere. Retool any places where there are voids.

Tool the head joints first.

Use a smooth, straight motion to tool the bed joints so that there are no irregularities.

The completed corner—tooled, brushed, and ready to continue with the brick structure.

Make sure that you allow the tooled mortar joints enough time to dry on the surface so they will not smear. When you have finished tooling the mortar joints, brush the wall with a medium stiff brush. I always use a long-handled brush so I don't skin my knuckles on the brick. The brushing provides the finishing touch to a completed brick project and makes the task of cleaning the brickwork later a lot easier.

In this chapter I walked you through the basics of bricklaying skills. These skills and techniques remain much the same regardless of what type of brick project you undertake. In Chapter 5, I expand on these techniques by explaining laying brick to the line, and I include a few interesting projects in which you can apply these skills.

Chapter 5

LAYING BRICK TO THE LINE

Without question, the most productive and accurate method of building any brick wall is to lay the brick to a line stretched tightly between two leads or corners. Even a professional mason can't build a straight wall without using a line as a guide. In Chapter 4 I explained the basics of building corners or leads because one cannot lay up a wall until the leads are built. In this chapter I describe proven techniques you should follow when laying brick to a line. I also point out some time-saving tips that should make your brick projects a lot easier and more enjoyable to build.

ESTABLISHING THE FIRST COURSE LAYOUT

Theoretically, a standard brick including the mortar head joint equals 8 in. in length. In reality you will discover that there is always a little variation in the length of brick, which has been previously mentioned. What this boils down to is that you never take a given measurement such as 24 ft. and expect it to work perfectly to 36 standard brick in length. It just won't happen!

Therefore, always dry-bond the first course of brick from one end to the other on the layout line, allowing ⅜-in. head joints, using the techniques for building a corner as described in Chapter 4. It may be necessary to open or close the head joints a little to work full brick over the length of the wall. If you don't do this, you could wind up with a small piece (usually 2 in.; called a *plug* in

Laying brick to a line helps ensure you'll build a straight wall.

Solving the Problem of a 2-Inch Piece in a Wall

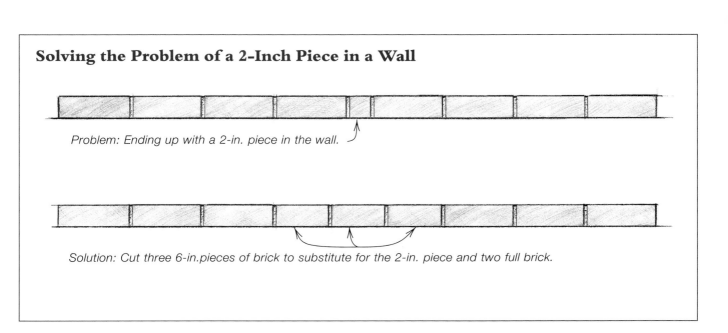

Problem: Ending up with a 2-in. piece in the wall.

Solution: Cut three 6-in. pieces of brick to substitute for the 2-in. piece and two full brick.

Dry-bond the first course of brick on the layout line, opening or closing the joints as needed to end with a full brick in the center.

Lay the first brick on the end to No. 6 on the modular scale rule.

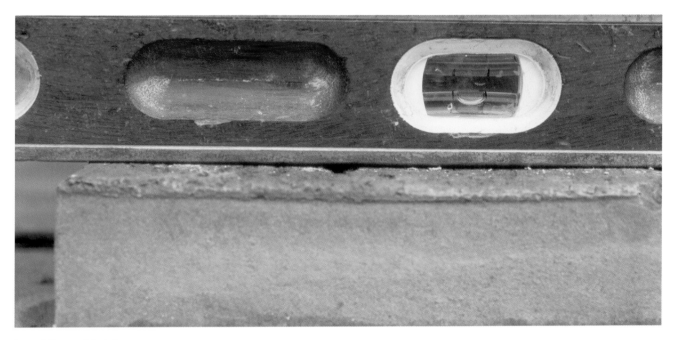

Level the end brick.

Plumb the face of the end brick.

the trade) or possibly a half brick (also known as a *bat* in the trade) in the center, which are big no-nos in the masonry trade. Sometimes a piece smaller than a full brick has to be used in a shorter wall, where you cannot change the measurements enough to suit full brick. In this event, always carry the piece against a door or window jamb, if possible. A good mason will never lay a cut brick less than 6 in. in the center of a wall. Sometimes the only way around this is to cut two or more three-quarter brick for the center to resolve the problem (see the drawing on p. 87).

After you have finished dry-bonding, lay one brick in mortar on each end of the wall to No. 6 on the modular scale rule (in this example). Be sure to take your measurement on the end or on the corner. Level the brick from the end out, and plumb the face of each end brick.

Attaching Line Blocks to the Ends

Attach a nylon line and line block to the end brick, stretching the line tight. To place a line block, first tie a knot in the end of the line, and slip it into the groove in one block so that it doesn't pull loose (see the top photo

Tie a knot in the end of the nylon line, and insert the line in the groove of the line block.

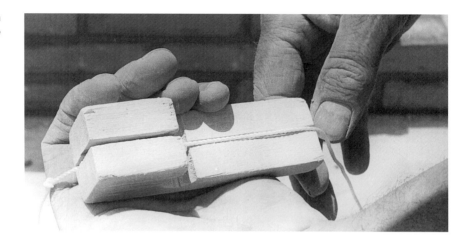

Wrap the line around the other block so it doesn't pull loose and attach the block to an end brick.

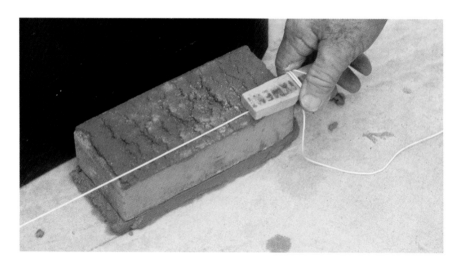

A typical example of the line attached to a line block on a brick corner.

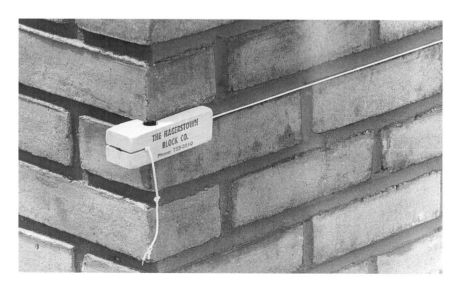

on the facing page). Hook the block with the line onto the opposite end of the wall even with the course to be laid (see the center photo on the facing page). Pull the line tight, wrapping the line a couple of times around the other block so it does not pull loose, and hook it over the end of the brick on the opposite end. Place a half brick on top of the line, and adjust it to make sure that the line is even with the top face of the brick. Be careful not to pull the line so tightly that it pulls the end brick loose.

Using a Line Pin and Nail to Hold the Line

Sometimes it is not practical to use line blocks. For example, if one person is building the corner and another is working on the wall at the same time, line blocks will interfere with building the corner. You can use a steel line pin and nail driven in the mortar head joints to accomplish the same thing. I recommend using a long nail (3-in. 10d common or larger) on the nail end, so it won't pull out of the mortar head joint. Not many things are more dangerous than a flying nail that has pulled out of the wall under pressure!

When building a brick wall, a line pin and nail are attached to the wall in the following manner. The nail end is called the *peg end*. Slip the line around the nail and tie a knot so that it will not come loose. Drive the nail into the mortar head joint on a slight downward angle so that the line will be even with the top of the course of brick that you are laying. Tug on it after you have driven it in to make sure it will not come out.

On the opposite end of the wall, drive the steel line pin into the mortar head joint on an angle so that the top of the pin is even with the top of the course of brick that you will be laying. Line pins can have several grooves in the edge that the line fits into. Pull tightly and wrap the line around the pin in the groove. Adjust the line so that it is even with the top of the course of brick. As a rule, the person working on the right end pulls the line. If working by yourself, you'll have to do both.

When the line is moved up for another course, be sure to fill in the holes left by the line pin and nail while the mortar is still fresh. If you wait until the project is completed before filling the holes, the mortar color may be a little different.

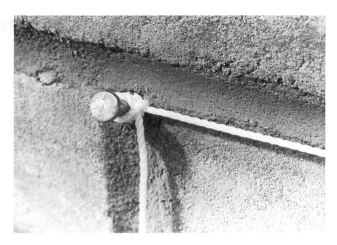

Securely position the nail end of the line in the head joint. (Photo © Delmar Publishers.)

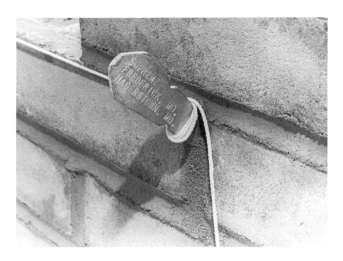

Pull the line taut, and wrap it around the line pin in the mortar joint. (Photo © Delmar Publishers.)

Laying the First Course in Mortar

Using a pencil, mark where the end of each of the dry-bonded brick is on the base, pull the brick off the line, and lay the first course of brick in mortar from each end to the middle of the wall (see the top photo on p. 92).

The last brick laid in the center is known as the *closure brick*. To decrease the possibility of the mortar head joints leaking next to the closure brick, I recommend

Lay the first course of brick to the line, working from the ends to the middle.

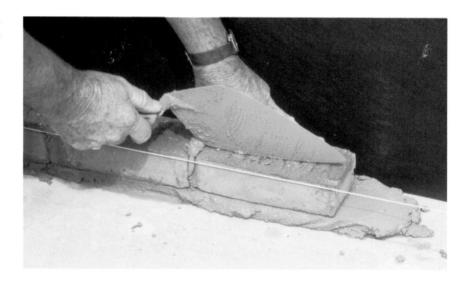

Before laying the closure brick, apply a head joint to each end of the brick in place. (Photo © Delmar Publishers.)

Apply mortar on each end of the closure brick, and lay it in place.

Keep the blade of the trowel well back from the line, and spread the mortar on the wall for only a few brick at a time.

Furrow the bed joint to even out the mortar before laying the brick.

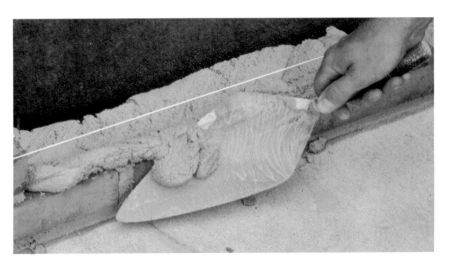

Trim any excess mortar from the bed joint.

Press the brick into position, using the trowel blade to nudge it to the line, if necessary. (Photo © Delmar Publishers.)

Apply the head joint for the next brick.

The top view of a laid brick shows that it is the correct distance ($\frac{1}{16}$ in.) back from the line. (Photo © Delmar Publishers.)

Build the lead to its predetermined height.

buttering a mortar joint on each end of the brick adjoining it and also applying a head joint to each end of the closure brick. This is known as *double jointing* and is very effective in obtaining a well-filled mort ar joint (see the center and bottom photos on p. 92).

Build the corner up to its predetermined height, as described in Chapter 4. After raising the line for another course, spread the mortar on the wall, being careful not to hit or cut the line. The blade of the trowel should be well back from the line during the spreading process (see the top photo on p. 93).

Until you get the hang of it, don't spread mortar for more than two or three brick at a time. Furrow the mortar bed joint with the point of the trowel, keeping the trowel blade back away from the line as much as possible (see the center photo on p. 93). Cut off the excess mortar with the front of the trowel blade (see the bottom photo on p. 93), staying clear of the line, return it to the mortarboard. Lay the brick in the mortar bed joint, pressing them down to the line as close as possible. It may take a little tapping or adjusting with the trowel blade to settle the brick into their proper positions (see the top photo on the facing page). Cut off the excess mortar, and apply it to the end of the brick just

laid to form the joint for the next brick (see the center photo on the facing page).

The proper distance for a brick to be laid back from the line is approximately ¹⁄₁₆ in.—just so you can see a little daylight between the line and the brick. This is not something you measure—you will readily perfect this skill after laying a few brick to the line. When a brick is positioned, the bottom edge should be flush with the face of the brick below and the top edge should be level with and ¹⁄₁₆ in. back from the line (see the bottom photo on the facing page). When brick are against the line, it is called *crowding,* and when brick are too far back, they are said to be *slack*. Neither is preferred if a plumb wall is to be built. Granted, it is impossible to lay brick to a line without disturbing the line to some degree, but with the correct techniques and practice disturbances can be held to a minimum.

TOOLING THE MORTAR JOINTS

When laying brick to a line, I recommend tooling the mortar joints about every three courses, so they don't dry out too much. A sledrunner jointer will do a much better job on a wall than the smaller pocket-size jointer because the joints are longer and will be straighter with-

Tool the head joints with a sledrunner jointer. (Photo © Delmar Publishers.)

Brush the wall and repoint any holes that are visible. (Photo © Delmar Publishers.)

Tool the bed joints with a sledrunner jointer.

out any dipping up and down. Always tool the head joints first. After all of the head joints have been tooled, run the sledrunner horizontally through the bed joints. Complete the tooling process by brushing the wall with a long-handled brush, repointing any holes that are left.

SETTING A TRIG BRICK

When building a long brick wall, it may be necessary to set a brick in the middle to the correct course height and fasten the line to it to prevent the line from sagging or being blown about by the wind. This brick is known as a *trig brick*. A trig brick is also used to keep the line stable if a number of masons are laying brick on the wall at one time. Generally, a brick wall longer than 32 ft. should have a trig brick set in the center.

One way to tell if a line is sagging in the middle is to put the line up for a course of brick on each end and sight down the wall from one end to the other. A sag can be readily seen. The other way is to nail a wood block in the center of the wall with the top at the same level as the bottom of the first course of brick at each end. Set the course rod on the wood block that was used as a gauge to build the corners, and lay the trig brick to the correct course height marked on the rod.

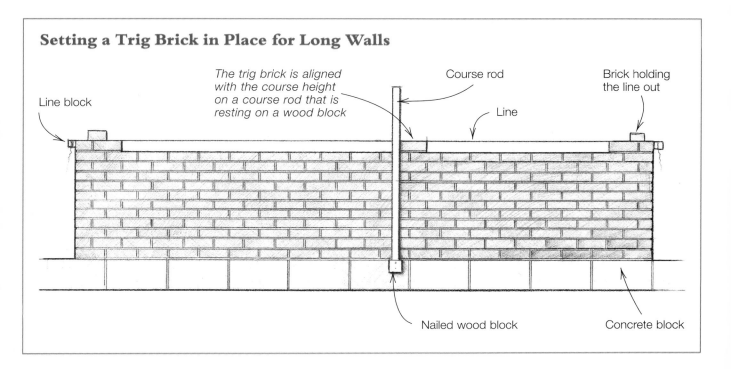

Setting a Trig Brick in Place for Long Walls

Line block

The trig brick is aligned with the course height on a course rod that is resting on a wood block

Course rod

Line

Brick holding the line out

Nailed wood block

Concrete block

In addition to being set to the proper height, the trig brick should also be level and plumb and should match the bond pattern of the wall. It is a good idea to sight down the wall every three or four courses to make sure the course you're laying is not getting out of line in relation to the trig brick. When sighting across the wall, have one person set the trig brick, snap the metal trig (a metal trig was shown in Chapter 2) on the line, and hold it even with the top of the trig brick. The other person can then sight down the line and determine if it is set to the same height and if it is still in line. This sighting method is accurate within ¼ in. Sighting brick courses should always be done when building long walls as the coursing may not always be exactly aligned with the scale rule, due to variation in bed joint thickness.

After the trig brick is correctly set, place a dry brick on top of the metal trig to hold the line even with the top outside edge of the brick while the wall is being filled in. A trig brick set incorrectly can cause a wall to be out of line or at the wrong height, even though the corners may be perfectly built.

A trig brick set into position to the line keeps the line from sagging or from being blown around by the wind.

Cutting Brick

In the course of laying brick, it is necessary at times to cut brick at openings such as windows and doors. They should be cut accurately and neatly for a good appearance. Although a masonry saw can be used to cut intricate or very thin pieces, for speed most cuts can be made with a brick hammer or a brick set chisel.

Using a Brick Hammer

When cutting brick with the brick hammer, start by marking where you want to cut on the face side of the brick, using a pencil and small square. Hold the brick in your hand with the face side up, cupping it slightly so it does not lay flat against the palm of your hand. (The reason for this is that sometimes when striking the brick with the hammer, it will momentarily open up and then close again without breaking cleanly, pinching the palm of your hand. This can cause a nasty blood blister to occur. It's a good idea to wear gloves when doing this.) Strike the brick along the pencil line with the chisel peen end of the hammer until it is scored but not broken through.

Turn the brick over in your hand and score the other three sides the same way. Don't strike the brick hard enough to cause it to break yet, only score it on the cut line with a light tapping motion.

When you have scored all the way around the brick, lay the brick flat in your hand on its widest side and strike it in the center with the blade of the hammer. It should break cleanly along the scored line. If the brick is stubborn and does not break, rescore and try again. It can't stand this forever!

Since the cut edge will be laid in mortar against the end of another brick, use the hammer to trim off any excess to allow a neat fit when laid in the wall. Hold your hand well under the bottom of the brick to prevent hitting it with the cutting end of the hammer. Wear a pair of gloves and safety glasses to protect your hands and eyes.

Using a Brick Set Chisel

Cutting brick with a brick set chisel makes a very neat, accurate cut and is highly desirable when you do not want any chipped marks near the cut edge or when you want a very precise cut, such as for fireplaces or arches.

Mark and lay the brick face up on a piece of wood to cushion it. Place the flat edge of the beveled blade of the chisel on the mark toward the finished cut edge you want. Strike the head of the chisel with a sharp blow from the hammer. It may take several blows to break the brick along the cut line. Trim the cut

Mark the brick where you want to cut it with a pencil and square.

Score the brick along the marked line with a brick hammer.

edge when finished. The beveled edge of the chisel causes the blade to move downward at a slight angle reducing the amount of trimming needed to finish the cut.

If you need to make a number of cuts the same size, lay out a row of brick and mark them all with a straightedge at the same time. This reduces the chance of size variations in your cut brick.

Marking Brick with a Straightedge

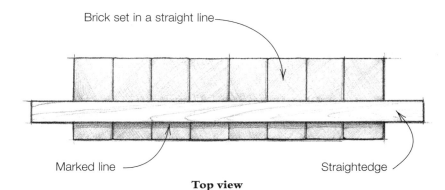

Brick set in a straight line

Marked line

Straightedge

Top view

Score the brick all the way around.

Strike the brick in the center to break it.

Trim the cut edge of the brick so that it fits neatly into the wall.

Cutting a brick with the brick set chisel is accurate and neat.

APPLYING A FULL MORTAR JOINT ACROSS THE LONGEST SIDE OF A BRICK

Up to this point, I have only discussed the normal mortar head and bed joints. In the course of building different brick projects, there will be many occasions when a mortar joint has to be applied across the longest side of a brick. Examples of this are when laying a header course, putting in brick windowsills, laying soldier courses over windows, capping off a wall, or doing arch work. Technically, this is still a head joint but many masons call this a *cross joint*.

It is applied in the following manner. Hold the brick in your hand by its longest dimension. Pick up a trowel of mortar, snapping your wrist slightly to set it on the blade, and swipe it down the side of the brick with a little force to make it stick to the edge.

Swiping a cross joint on the longest side of a brick should be done in two movements once you get the hang of it.

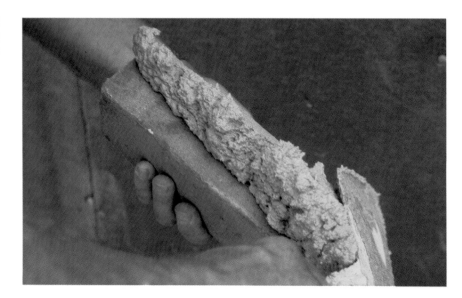

Complete the cross joint on the long side of the brick. (Photo © Delmar Publishers.)

Corner Poles

Metal *corner poles* (also called *masonry guides*) can be rented or purchased through masonry suppliers and substitute for building a brick corner the traditional way. They are used when a lot of brick walls have to be built, such as on a house or a commercial building. They are great time-savers and speed up the work. Most are made of lightweight aluminum and can be attached to the wood frame of a house when doing brick veneer work.

They set up rather easily. One end usually has a flat metal plate that rests on the foundation or where the brick corner starts. The pole is plumbed and the top end is fastened to the framework and secured with an adapter fixture. Adjustable steel tapes are used on the pole to fit either of the brick coursing scales for brickwork. Regular line blocks are attached to the pole at the correct height to hold the line. As the work progresses, the block slide up the pole for each succeeding course. The average corner pole is about 8 ft. high and can be moved up and reset if the building is more than one story.

A corner pole is set up on a brick veneer house to help speed up the brickwork.

Pick up another trowel of mortar the same way, and starting at the back of the brick, swipe it on the edge in the opposite direction. If the mortar is too stiff, add some water so it adheres to the brick. Practice this until you get the hang of it. The idea is to be able to butter a full joint of mortar on the brick using only two movements.

REFACING A BRICK PORCH AND STEPS

Here is a typical example of a masonry renovation project that can improve many older homes. When the house in this example was originally built, the porch and steps were constructed of concrete block and concrete. Over the years the porch settled a little and cracked in places, seriously marring the appearance of the front of the home. Since the original porch and steps have long since stopped settling, a practical and economical solution is to reface or veneer them with brick. There is no reason to go to the expense of tearing the entire porch down and starting over again.

This project combines building corners, laying brick to a line, and installing a set of brick steps. It utilizes many of the skills and techniques discussed so far in the book. The end result transforms any ugly old porch and steps into a pleasing entranceway at a minimum cost and increases the value of a home. For those who own rental properties, it is an excellent way to add a facelift to the front of a house inexpensively and increase rental appeal to prospective tenants. The biggest investment is your time.

Pouring a Concrete Footing

The first step is to excavate the earth and pour a concrete footing around the outside edges of the porch and steps to support the brick veneer. In the example shown in the top photo on p. 102, I plumbed down with the 4-ft. level from the outside edge of the projected concrete porch top to ground level. You can sprinkle a little white lime powder or sand around this line to mark it. The reason you do this is that the face of the new brick

Plumb the edge of the existing concrete platform to the ground level.

Pour a concrete footing in the trench around the porch and steps.

wall must extend out about 4½ in. to allow the brick facing to hide the old concrete slab.

Excavate about 8 in. wide and 20 in. deep around the perimeter of the porch, or below the frost line in your area. Pour 6 in. to 8 in. of concrete in the trench, leveling it off with a garden rake. In this example, I laid bricks below grade in mortar to allow for any ground settlement. I did not need to pour any concrete in front of the steps as there was a 5-in. concrete sidewalk already in place that the brick could be laid on (see the photo below). This is the perfect place to utilize bags of preblended concrete because it will not take very much.

Laying the Brickwork Level with the First Step

I measured down from the top of the lowest existing concrete step to the concrete footing with my scale rule and built up the brickwork that was below grade so that it worked out even with the top of the concrete step. Any of the brick that are laid below grade can be chipped or old ones. The front of the concrete step was two courses of brick lower than the grade on the sides of the porch, so I used face brick since they would be exposed (see the top photo on the facing page).

Build the brickwork on the front and sides of the steps level with the top of the first concrete step.

Build the porch walls and the ends of the steps, making sure the courses work out level with the existing porch concrete slab.

Building the Sides of the Porch and the Ends of the Steps

Using the scale rule as a guide for the coursing, I laid the brick on the sides of the porch and over to where the steps started, using the level and a line. I made sure that the brick courses worked out evenly with the top of the existing concrete porch slab (see the bottom photo above). To achieve this, the last course of brick on the porch wall worked out to be a rowlock. As I built up to the height of each individual step, I laid rowlock brick to form the end of the steps. Each step consisted of stretcher brick with a rowlock laid on top of it to an approximate height of 7 in., which is standard for a brick step. The first step was a little higher due to the uneven height of the sidewalk.

Start to lay the rowlock course of brick on the step to the line.

I stretched a line at the top of the steps taut from corner to corner to serve as a guide for laying the brick rowlock course. I blocked the line out on each end so it was in line with the top outside edge of the steps. To prevent getting any bumps in the steps and to keep the back edge of the brick rowlock course level, I used a long level and tapped the back ends of any high brick into position (see the photo above). I used a standard ⅜-in. mortar head joint and worked from both ends to the center.

About every fourth brick, I checked with the modular scale rule to make sure that I could work a full brick into the center of the step (see the photo at right). When I was several brick away from the last rowlock brick in the center, I made room for one extra head joint, as I would need a head joint on each side of the last brick laid in the step.

When the steps were completed, I laid a brick header course to the line that bonded across the porch wall and on top of the existing concrete platform around the porch (see the top photo on the facing page). I projected them out approximately ½ in. from the face of the

Check the spacing of the rowlock brick with the modular rule to No. 6 on the scale to be sure it will work out to a full brick in the center of the step.

Lay the brick header border course on the existing porch concrete slab.

The completed border course.

wall. This allowed the water to drip off and not run down the face of the wall. The header course would serve as a border for the brick to be laid on top of the porch. I completed the top of the porch by laying the paving brick to a line stretched across the borders. If you have a situation where the existing landing is too close to the door opening to allow for the thickness of a brick veneer, I would use half-size split brick, which are available from most brick suppliers.

Completing the Job

After the mortar joints cured for seven days, I cleaned the brickwork by washing it down with a solution of 1 part muriatic acid to 10 parts water. I waited another day for the area to dry out, laid a few brick on their ends for a flower bed, regraded around the porch, and planted some shrubbery to finish the job.

The completed refaced brick porch and steps.

LAYING A BRICK WALK DRY IN STONE SCREENINGS AND SAND

Here is another very useful and rewarding project you can build around the home, which will be appreciated by everyone. By not using mortar to lay the brick, the amounts of time, effort, and cost for this project are greatly reduced. The principal advantage of laying a brick walk in a dry bed is that it allows the brickwork to move with temperature changes throughout the seasons of the year without having to worry about the possibility of mortar joints cracking and needing repointing. Even though the brick are not laid in mortar, you still need a line to lay out the forms and the brick border around the walk.

Preparing the Base

You cannot just start by laying the brick on the bare ground. A proper base has to be prepared for the walk, even though it is being laid in a dry mix. The first step is to decide how wide you need to excavate the base. A 36-in.-wide walk is a good all-around choice. Using a standard paving brick, which is about 7⅝ in. long, lay out the brick dry on top of the ground in the pattern you want, including a border of brick on each side.

I selected a basket-weave pattern for the walk, which is shown in the photographs on pp. 108 and 109. To estimate the number of paving brick needed, calculate the square feet in the walk and multiply this figure by 4.5. I allow 5 percent for waste to cover any cracked or broken brick. Try to eliminate small pieces and work full brick as much as possible. Don't forget to allow enough room to install wood forms along the borders.

Drive a wood stake at each end of the walk, and stretch a line between them to use as a guideline for the excavation. It helps if this line is level because you can determine how much fall or pitch to allow by measuring down from the line. (This allows water to drain from the completed walk.) An easy way to do this is to hook a small metal line level on the line and adjust the line until it is level. Excavate the base approximately 6¼ in. deep and level it off with a rake. This allows 2¼ in. for the brick, 3 in. for fine gravel or stone screenings, and about 1 in. of sand to act as a cushion on top of the screenings.

Set the 2x4 wood forms in place for the brick walkway.

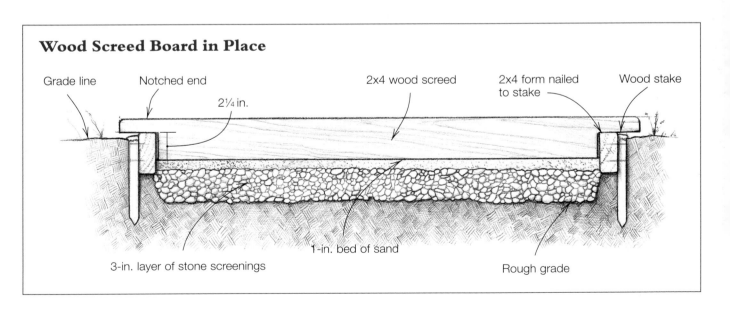

Wood Screed Board in Place

Grade line

Notched end

2¼ in.

2x4 wood screed

2x4 form nailed to stake

Wood stake

3-in. layer of stone screenings

1-in. bed of sand

Rough grade

Screed the fine stone screenings and sand so that the bed is even and ready for laying the brick.

Settle the brick into the sand bed, using a paving bond such as the basket-weave pattern shown here.

Installing the Border Forms

Drive wood stakes in the ground, and nail the 2x4 wood forms to them in the proper position and height (as shown in the photo on p. 107). I set the side of the forms on the inside of the walk about ¾ in. higher than the opposite side to allow water to drain off the finished brickwork.

Filling in with the Stone Screenings and Sand Base

You can simplify the task of placing the fine stone screenings and sand to the proper height by making up a wood screed out of a piece of 2x4 and dragging it across the top of the forms. Make the cut-out area in the screed the same height as a brick to create a near perfect bed to lay the paving brick on.

Laying the Paving Brick

Now comes the easy part, except for some aching knees later. Starting at one corner of the walk, lay the border brick on their edges snug against the edge of the 2x4. Next, start laying brick in the basket-weave pattern, two in one direction and two in the opposite direction. Settle each brick firmly into the sand bed by tapping lightly on top with the end of the hammer handle. If you crack a brick, discard it as the crack will only get bigger later. If you run into some low spots, simply add a little more sand and re-lay the brick. Periodically, check the ends of the pattern by squaring off from the border as you progress to make sure it remains square and in line. It also helps to lay a straight piece of 2x4 on its edge across the walk every couple of courses from border to border to check for any high or low spots.

Sweeping Sand in the Joints

After all of the brick have been laid, complete the job by sweeping sand into the brick joints. Eventually rain beating on the walk will wash out some of the sand. It is a simple task to resweep more sand in at a later time. Outside of spraying some weed killer on grass that is bound to grow between the joints, there is no other upkeep for a walk of this type. (A good way to prevent grass growth between the brick paving units is to install a moisture-resistant barrier such as black plastic between the sand and the stone screenings layers. It can also be

Sweeping sand in the joints between the brick is the final step.

positioned on top of the sand layer.) After you remove the form boards, replace the soil around the walk, and tamp tightly.

In this chapter, I have expanded on bricklaying techniques and skills with special emphasis on laying brick to the line, which is how the bulk of brickwork is done. The two projects present opportunities to utilize the information presented throughout the chapter. They are only a sampling of the many brick projects one can build. In the next chapter I explain how concrete block are laid and include some projects that are built of block and brick.

Chapter 6
LAYING CONCRETE BLOCK

Techniques and skills that are performed when laying concrete block are somewhat different from those used for laying brick. There are only several basic bonds used for block work, which are the same ones discussed for brickwork in Chapter 3. The most frequently used types and sizes of block available were discussed in Chapter 1. In this chapter I explain how different size block corners are started, how to plan and lay out a concrete block project, and how to estimate concrete block and mortar. I point out tips and techniques used when laying concrete block and cover the more important elements of building a successful concrete block foundation.

Because concrete block are larger, cost less per square foot, and go up much faster than brick, they are a logical choice for large commercial-type buildings and masonry structures around the home, such as foundations, garages, and tool and garden sheds. An interesting comparison between brick and block is that one 8-in. x 8-in. x 16-in. concrete block will occupy the same total area as 12 standard brick. As you can readily see, using block can save considerable time as long as concrete block suit the purpose at hand. Today almost all foundations are built of either concrete block or poured reinforced concrete. Used together, brick and concrete block form a strong, economical, long-lasting, and maintenance-free masonry wall.

Many of the techniques and skills required for laying concrete block in a foundation are the same as those for laying brick.

PLANNING AHEAD

You should think ahead when designing a block project and try to make it work with as few cuts as possible. It not only makes the wall look more professional, but it also saves time, materials, and wasted effort. I think in multiples of 8 in. as much as possible because the majority of block are 8 in. high and 16 in. long. The same is true for the height of a block project. The window, door, and wall heights should all work full block wherever possible. There are, of course, some exceptions, but in most situations, the number of cut pieces can be limited.

The simplest method I have found to achieve this is to draw the project first on a sheet of ¼-in. graph paper, which is available at most office or art supply stores. Use a scale that fits the size of your project. You don't need any professional drafting experience to do this. A practical scale that I use frequently for brick or block work is ¼ in. represents 4 in. Since windows and doors are usually modular in size and built to standard heights, they generally do work out to full block.

The drawing on p. 112 shows a typical plan layout view of a tool or garden shed. The foundation measures 10 ft. wide x 14 ft. long, which works full-size block. The elevation side view of the tool shed utilizes the same mul-

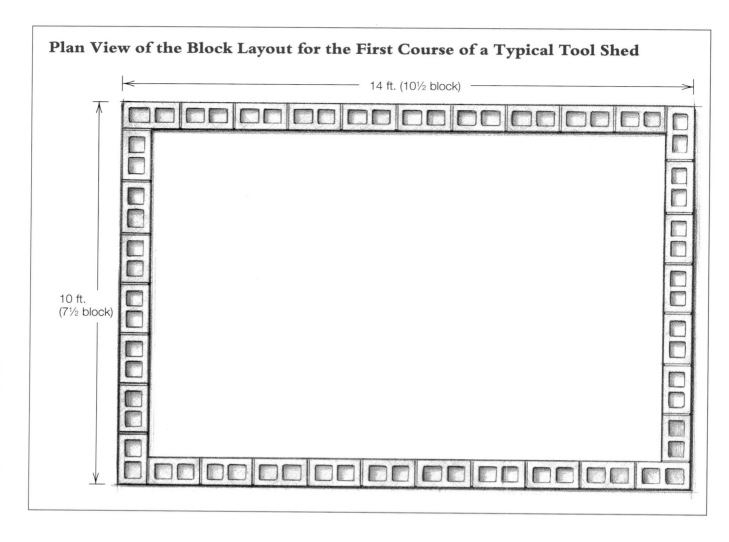

Plan View of the Block Layout for the First Course of a Typical Tool Shed

14 ft. (10½ block)

10 ft.
(7½ block)

tiples of 8 in. (see the drawing on the facing page). Notice that the door and window heads are on the same level. The block courses also work out evenly to the sill under the window and from the window head to the top of the last course. Because a standard door width is 36 in. and the window width in the example is also 36 in., 4-in. and 12-in. cuts have to be made against the jambs. Windowsills work out as necessary by cutting the block under the window or by using concrete brick. Although the example used here is a relatively small building, you can readily see the advantage of laying out the block to fit into a scale, regardless of the size of the structure.

LAYING OUT THE BOND

Because concrete block are consistent in length, a wall can be measured out from one end to the other without dry bonding. There are several methods of marking the bond off on the base. One uses the 6-ft. mason's folding rule to mark individual block in 16-in. intervals. Another method commonly used by bricklayers on the job uses a 4-ft. level in place of the rule. Because 48 in. is three times 16 in., you simply keep marking at the end of the level, sliding it down the base, and re-marking until the opposite corner is reached. This method is a little hit-or-miss on a longer wall—if you don't match up each mark accurately when you shift the level, it can get off some.

Side View of a Block Tool Shed Wall

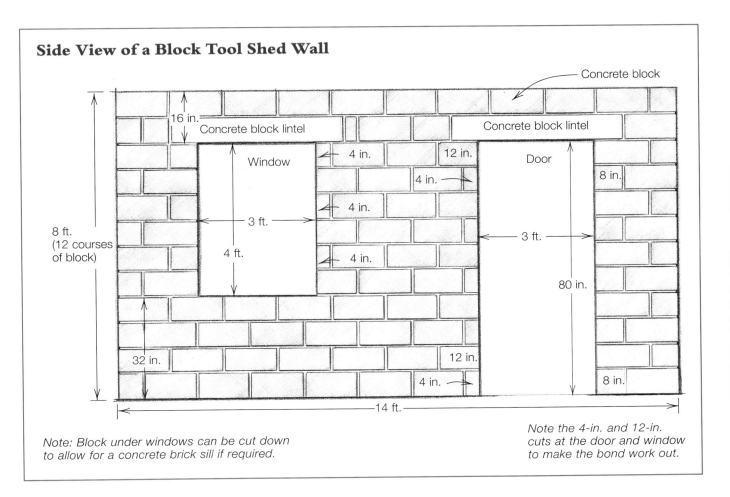

Concrete block

16 in.

Concrete block lintel

Window

4 in.

12 in.

Concrete block lintel

Door

8 in.

4 in.

8 ft.
(12 courses
of block)

3 ft.

4 in.

4 ft.

4 in.

3 ft.

80 in.

32 in.

12 in.

4 in.

8 in.

14 ft.

Note: Block under windows can be cut down to allow for a concrete brick sill if required.

Note the 4-in. and 12-in. cuts at the door and window to make the bond work out.

The third method uses a steel measuring tape; I recommend it because it is the most accurate. Start by driving a nail in the concrete footing at one end of the wall. On the end of the corner, stretch the tape out tightly and mark every 16 in. with a yellow lumber crayon or pencil on the base. Almost all steel tapes I have used have diamonds or red marks at these 16-in. intervals, which simplifies the task. On small projects, I use the folding rule or level, and on longer walls such as a house foundation, I use the steel measuring tape.

STARTING A CORNER

The starter size of a block on the corner depends on the width of the block being laid. As previously mentioned, the only block that will bond over the one beneath it

without cutting or adapting is the 8-in. block (8 in. is half of 16 in.). There are specially made L-shaped corner starter block for 10-in. and 12-in. foundation walls that will produce the required 8-in. lap. This was previously discussed in Chapter 1.

Three popular block widths that require a starter piece on the corner are the 3-in. block, the 4-in. block, and the 6-in. block. The way you determine the length of the starter piece is to add the width of the block being laid to 8 in., which is the standard lap over each one when laid in the wall. For example, if you are going to build a 3-in. wall, you need an 11-in. corner piece. If building a 4-in. block wall, you need a 12-in. corner piece, and if you are going to build a 6-in. wall, you will

Corner Layouts

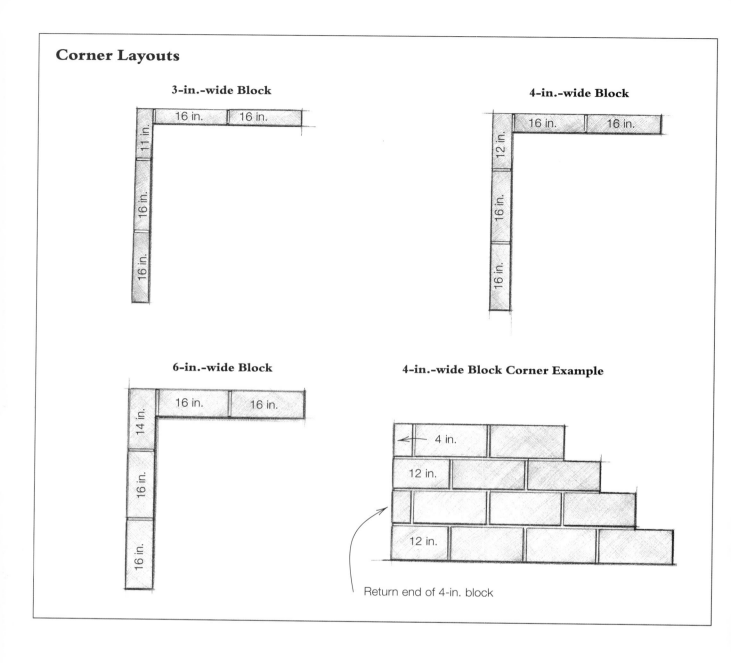

3-in.-wide Block

16 in.   16 in.

11 in.

16 in.

16 in.

4-in.-wide Block

16 in.   16 in.

12 in.

16 in.

16 in.

6-in.-wide Block

16 in.   16 in.

14 in.

16 in.

16 in.

4-in.-wide Block Corner Example

4 in.

12 in.

12 in.

Return end of 4-in. block

need a 14-in. corner piece. It is possible to buy a 4-in. block that is 12 in. long at some block suppliers, but I do not know of anyone who makes starter block for 3-in. or 6-in. walls. It is a lot simpler just to cut the starter piece as needed, rather than worry about trying to special-order them. The drawing above illustrates the proper corner pieces for the 3-in, 4-in., and 6-in. block and a 4-in. block corner.

TECHNIQUES OF LAYING CONCRETE BLOCK

As I have stated previously, concrete block techniques are somewhat different than the techniques for laying brick because block are larger and heavier. Because they are so much heavier, they have to be laid in a mortar joint with some finesse and not just dropped in the mortar bed joint, as they will surely sink too low and require re-laying.

Laying Concrete Block Correctly

A concrete block has a definite top and bottom. It is easy to tell the difference if you know what to look for. The top of a block has slightly wider edges and webs, including the interior cells. There is a very good reason for this—it is a lot easier to spread mortar on top of wider edges, and it is much less likely that the mortar will fall down into the cells. This may seem like a trivial point, but it makes a big difference in the block being supported on the mortar bed properly. Laying block right side up increases the numbers of block anyone can lay in a given period of time. However, having a block laid upside down doesn't affect the structural strength of the wall.

Laying the First Course

The block wall shown in the following example is a short panel wall I built to illustrate proper techniques. The bond should be laid out to work whole block before spreading any mortar, allowing ⅜ in. between each block for head joints. The mortar bed joint under the first course should be completely filled and solid to prevent water leaking under it and to provide maximum support.

The block on the left shows the bottom of the block and the block on the right shows the top. Notice how much wider the center web is on the block on the right.

Form a full bed joint to prevent water leakage.

Level the block crosswise only on the first course. (Photo © Delmar Publishers.)

Estimating Concrete Block, Mortar, and Sand

A simple rule-of-thumb estimating procedure is more than adequate for small or average size jobs up to a house. As in estimating brick, large concrete block jobs are estimated from specific tables with the use of computer-estimating programs and are computed by trained estimators. You will find that the methods I describe here are simple, quick, easy to use, and accurate. They are based on many years of practical experience and are used by many masonry contractors.

Estimating Block

The easiest method of figuring how many block are needed for your project is first to add all of the wall lengths of the project together to determine the total linear length. Since a standard block is 16 in. long, there are three block to every 4 ft. in length. This is a 3:4 ratio. Multiply the total linear feet by 0.75, and the result will be the total number of concrete block on one individual course. For example, let's use a regular block garage to illustrate how this works. The garage is going to be 20 ft. by 24 ft. Add 24 to 24, which equals 48, and add 20 to 20, which equals 40. Adding 48 and 40 equals 88 linear ft. Multiply this figure by 0.75 and the answer is 66 block on a course. If you have any openings, such as doors or windows, deduct for them.

Next, assuming each course of block is 8 in. high including the mortar bed joints, divide 8 in. into the total height of the garage (also in inches). This gives you the number of courses of block high. Then multiply the number of block on one course by the total of number of courses high. Now you have the total number of block needed to build the garage. Referring to the above example, let's assume that the garage height is 10 ft. including the foundation. Change this into inches by multiplying 10 ft. by 12 in., which equals 120 in. Now, divide 8 in. (the height of one block) into 120 in. This equals 15 courses of block. Multiply 66 (block on one course) by 15 (number of courses high of garage), which equals 990 concrete block. Always allow about 5% extra blocks for waste or broken blocks.

If you are building a 10-in. or 12-in. foundation block wall, remember to allow one special L-shaped corner block for every course on each corner. I always figure the total number of block needed and then deduct the number of L-shaped corner

Because the top of a concrete block is made square with the face, you can level across the top at its narrowest and the block will also be in a plumb position. This technique should only be used on the first layout course. From this point on, the face of the block wall is plumbed vertically, the same as in brickwork.

Then, level the block horizontally. Don't settle it entirely into position until the course height has been checked with a mason's spacing rule. Lay a concrete block at the opposite end of the wall, allowing for three block in between and making sure it is level with the first block laid. Since three block require 48⅜ in. including the closure head joint, this is a shade longer than the length of the 4-ft. level. Extend the trowel slightly out over the edge of end of one block, and rest the end of the level on it to compensate for this.

Applying the Head Joints

Attach a line to the wall with a line block, and lay the first course to the line, as described in Chapter 5. Applying a mortar head joint to the ends of a concrete block is different than to brick, but it is still relatively simple. Pick up half a trowel of mortar, set it on the trowel blade with a quick downward snap of the wrist, and then swipe it on the end of the block. Using a slight pinching motion with the trowel blade, press the mortar out a little so it does not fall off when lifting and laying block. Using the same example, you know the garage has 15 courses. Since there are four corners on the foundation, multiply 15 by 4, which equals 60 L-shaped corner block. Deduct 60 from 990 block, which equals 930. So you need 930 regular concrete block plus 60 L-shaped corner block. You do not have to order square end block for regular corners, as most block are square on the ends anyway.

Estimating Mortar and Sand
If you are using masonry cement mortar, which most masons do, figure that 1 bag when mixed will lay about 30 block. This also allows for about a 5 percent waste factor, which is normal. Referring to the example cited, divide 30 into 990 concrete block, which equals 33 bags of masonry cement.

To estimate the amount of sand, figure that it takes 1 ton of sand for every 8 bags of masonry cement. Divide the total number of bags of masonry cement, which for this example was 33, by 8, which equals 4.125. I always allow a little extra sand, so round this off to 4½ tons of sand.

If you are going to use a portland cement/lime mortar instead of masonry cement (based on 1 part portland cement, 1 part hydrated lime, and 6 parts sand), figure that 1 bag of portland cement to 1 bag of hydrated lime to 42 shovelfuls of sand will lay about 62 block.

If you are using the standard 40-pound bag of preblended mortar (which has cement and sand already in it), figure that 1 bag will do about 17 block. The

consistency and stiffness of mortar for laying concrete block is about the same as for laying brick, so the water content is the same.

A word of caution about mixing mortar for block: Masons usually mix 18 dirt shovelfuls of sand to one bag of masonry cement, which is called a batch. (This is a 1:3 ratio.) There is no way to pick up the same exact amount of sand on a shovel each time; however, try to be as consistent as possible. It does not have be perfect, but it should be reasonably close. Some masons prefer measuring mortar ingredients by using a standard 5-gallon drywall bucket as a gauge in place of a shovel. Either way works, depending on how accurate you want to be. Just remember to use the correct ratios.

Level the block horizontally.

Level from one end block to the other, using the trowel to make up the distance.

Swipe the head joint on the end of a block before laying the block in the wall. (Photo © Delmar Publishers.)

the block in the wall. A head joint can also be swiped on the end of a block once it is laid in the wall to receive the next block laid against it, but this takes a lot of practice.

Always butter the head joints on the ends of the last block to be laid on a course and the block already in place before laying the last block into position to the line, the same as laying a closure brick. This prevents the joint from leaking.

Applying Mortar

When laying succeeding courses, mortar is spread on top of the block differently than on brick. Generally, mortar is not spread across the center of the webs of a block unless specified for some special reason. It can be spread by flopping it off the trowel on the block, by swiping it on the edges, or by a combination of both. The simplest and most efficient method in my opinion is to pick up a modest trowel full, set it on the trowel blade with a gentle downward snap of the wrist, and swipe it on the outside edge of the block. Do not hold the trowel perfectly vertically or the mortar will tend to

Pinch the head joint on with the trowel blade so that the mortar will not fall off.

Lay the closure block with mortar on the ends of it and on the block already in place. (Photo © Delmar Publishers.)

fall down into the open cells. The correct position for the blade is on an angle with a little lean to the inside of the block. Swipe the mortar on with impact to make it stick to the edge, as shown in the bottom photo on this page. Reverse this motion for the opposite side of the block. It only takes a little practice to learn how to do this.

Laying Block to the Line

All masons tend to develop individual styles. There are, however, certain techniques I recommend that you follow when laying block to the line, which will make your block laying a lot more comfortable and less of a strain on your wrist and back. First of all, pick up the block at both ends for good balance, keeping your fingers in the center to avoid getting them in the mortar head joints. Don't get in the habit of laying the trowel down every time you lay a block and then picking it up again to adjust the block to the line. Instead, hold the trowel so that your fingers wrap around the handle and the block at the same time when lifting and laying it in the wall.

Swipe the mortar on the top of the block with the trowel, using impact to help the mortar stick to the edge. (Photo © Delmar Publishers.)

You will discover that this is relatively easy to do with a little practice. Keep your feet close together to relieve the stress on your back. Your legs should sustain most of the weight.

Lay the block into the mortar bed gently, and at the same time, press it back so the mortar head joint squeezes out, forming a well-filled joint. The trick is not to drop the block suddenly, but to lay it into the bed joint gradually so it will not sink below the line. I would compare it to the motion of slowly immersing a tea bag in a cup of water.

Adjust the top of the block to the line by tapping on the center of the end with the trowel blade to prevent chipping the face. The palm of the opposite hand should rest on the end of the previously laid block so that you can feel if they are level with each other. A brick ham-

Pick up a block at both ends to balance the strain on your wrists. Practice holding the trowel when you do this to save reaching for the trowel each time you lay a block.

The mortar head joint should squeeze out to fill any voids.

mer can be used to settle a stubborn block into place, but it shouldn't be necessary to position every block to the line.

Concrete block are laid approximately 1/16 in. back from the line—the same as brick. Always lay block from each end to the center of a wall in case they have to be adjusted to work full block. Don't move a block once the mortar has taken an initial set. If it has to be moved, it is best to take it up, spread fresh mortar, and re-lay it.

Cut off any protruding mortar holding the trowel blade at a slight angle to prevent smearing the face of the block. This angle also helps to keep from tearing the mortar off and from leaving holes or voids in the joints.

Tooling the Mortar Joints

I prefer using a sledrunner-type jointer for tooling block mortar joints as it forms a much straighter joint than the smaller pocket-size one without any dipping up and down. After allowing enough time for the mortar joints to surface dry (once they've reached thumbprint hard),

I tool all of the head joints first. Then I run the sledrunner through the bed joints, filling in any holes or voids as I go (see the top photo on p. 122).

You can cut down on smearing the joints by shaving off excess mortar from the edges of the tooled joints with the trowel blade before brushing the wall (see the bottom photo on p. 122). A word to the wise! Keep your fingers out of the mortar as much as possible. It is tempting to take a hunk of mortar on your fingers to fill a hole but there are many masons who get what is called *cement poisoning,* which causes their fingers to crack open and bleed. This is very painful, and only time and keeping your fingers out of the cement will cure it.

Cutting Concrete Block

In the course of laying block, it is inevitable that sometime you will have to cut. Block are more brittle than brick and are subject to premature cracking during the cutting process, if you're not careful. You can use an abrasive blade mounted on a circular saw or cut the tra-

Position the concrete block about 1/16 in. back from the line. (Photo © Delmar Publishers.)

Cut off the excess mortar from the bed joint, holding the trowel at a slight angle.

After tooling the head joints, tool the bed joints with a sledrunner jointer, forming smooth horizontal joints. (Photo © Delmar Publishers.)

Shave excess mortar off the face of the wall with the trowel after tooling the joints. (Photo © Delmar Publishers.)

Score the block with the brick set chisel and hammer before cutting it.

Finish the cut at the top edge of the block.

ditional way using a brick set chisel. If you are going to use a circular saw, be sure to wear safety glasses and a dust mask.

When cutting a block with a brick set chisel and hammer, follow this procedure. Mark the cut on both sides of the block with a pencil and square first. Lay the block on a relatively soft surface that will cushion the blows.

Old Mother Earth works fine for this. Score along the marks on each side of the block with the brick set chisel, striking hard enough to score it but not cause a complete break. Next, lay the block on its bed side, and strike it on the top edge in line with the scored edge. With any luck at all, it should break cleanly. Trim off the edges of the cut vertically by chipping lightly with the end of the hammer blade. This should remove any bumps on the edge (see the photo on p. 124).

Trim the edge of the cut with the brick hammer.

BUILDING CONCRETE BLOCK FOUNDATIONS

Concrete block are used more often to build foundations than any other masonry unit. While it is true that there is increased use of reinforced concrete for foundations, particularly in large developments where forms for the same house design can be used over and over again, concrete block are still very popular because of the cost, the ease of construction, the fact that no forms are required, and the availability of the product. Concrete block can be laid and built to fit almost any angle or shape, they are strong, they resist fire and moisture, and they have a very low maintenance cost.

While a foundation is one of the most important structural parts of any house or building, I have found in my experience that it is often the most neglected. In addition to being structurally sound, foundations also must be waterproof and properly drained. There is nothing that hurts the resale of a house more than a leaky, moldy, and musty basement!

Space does not allow me to cover all aspects of building a block foundation, but I would like to point out some of the more important details to be considered in building a typical concrete block foundation. There are two principal types of foundations: a crawl space where plumbing pipes, electrical conduits, and heating and cooling ducts are concealed, and a basement that is actually another usable floor of the house.

Generally the height of a crawl space should be at least 4 ft. I base this statement on crawling around in many crawl spaces with my nose in the dirt and my head banging overhanging pipes, never quite knowing what I might meet up with in the dark. The minimum finished height of a typical full-size basement should be from 7 ft. 6 in. to 8 ft. It can be higher if you want to keep pipes up out of the way.

Starting with a Good Footing

If you are going to build a crawl-space foundation, be sure to check the building codes in your area for the required depth below the frost line, and of course, you will need a building permit.

After the foundation area has been laid out with a transit level and excavated from the properly determined finished above-grade height, the concrete footings are poured or placed to the correct height for the block walls to be built on. The standard rule for concrete footings for a block wall is that they should be twice as wide as the block being laid and 8 in. deep. This meets all standard building codes I know about.

A concrete block foundation for a house that has a full-size basement should be built of either 10-in. or 12-in. block, depending on the amount of fill that will exert pressure against the exterior of the wall. Therefore, a footing for a 10-in. block wall should be 20 in. wide to spread the load and 24 in. for a 12-in. block wall. Specify footing grade concrete when you order from the concrete plant as it is a little less expensive than finished floor concrete. (Additional information on concrete work is covered in Chapter 7.)

Allow a day or two for the footing to harden. Reestablish the foundation corner points with the transit level and indicate them by driving nails into the concrete corner points. Then strike a chalk line between the nails indicating the outside of the walls.

Placing Block in the Foundation

You can save yourself a lot of back-breaking work by having the truck driver unload the block in the foundation area near where the walls will be built. Concrete block trucks now are equipped with a hydraulic boom that allows them to pick up and set the block inside the foundation wherever you want them.

This is one time you want to be certain to be on the job to make sure that the block are not put too close to the wall lines. I usually allow 2 ft. for working room. Keep in mind that scaffolding also must be erected to complete the foundation. Be sure to leave traffic lanes for the mortar wheelbarrow, so it can get around to stock the mortarboards near the walls. Generally, I leave one end of the foundation wall down as long as possible to move materials and equipment in and out. If the truck is going to drive across the footings into the foundation, it is a good idea to block up some heavy wood planks a little bit across the footings so the truck does not break them.

Laying Out the First Course

You will need an L-shaped starter block laid on each corner with a couple of regular block laid against it so that it doesn't pull loose when attaching the line. Determine the bond by taping it off with the steel tape. I recommend laying one course of block in mortar all of the way though the wall you intend to work on before building a corner to make sure that the bond will work and the wall is in line.

Sometimes there are steps in the footing due to a changing grade line or unstable earth. Always build up the lowest area first to the point where a continuous course

Lay out one course of block to make sure the bond works out correctly.

Lay the block level with the steps in the footing.

of block can be run from one corner to the other. Steps in footings should be in multiples of 8 in. so they will work full block.

Making a Course Rod

Before building the corners, make a course rod, figuring from the top of the foundation down in multiples of 8 in. Any special features such as beam pockets and windows should be marked on the pole. Each corner should be checked for height from a benchmark as it is built.

The corners can be laid up in the traditional manner or by setting up a metal corner pole (described in Chapter 3) and using it as a guide for the line. When tying in porches or garage walls that are higher than the main foundation, you can bridge the gap with a concrete block lintel. Set one end of lintel across a wall on the low end and let the other end rest on a concrete pad that has been placed at the higher level. This saves a lot of time and materials for areas that do not require a full-depth basement.

Lay off the block courses on a course rod, being sure to mark doors and windows.

Bridge the span between a garage wall and a house foundation with a concrete lintel to a concrete pad.

Set the block wall back for brick veneer once you hit finished grade.

Setting the Wall Back for Brick Veneer

As you build the block work up to the natural grade line of the earth, you should set the front of the wall back 4 in. to form a shelf for the brick veneer, which shows above grade. Seldom does the block work run all the way up to finished grade! I always allow an extra three courses of brick below grade to compensate for the earth

settling around the foundation. It looks very unprofessional to see a course of block with tar showing above finished grade. This is done by switching to a narrower width block at this point. For example, a 12-in. block wall would be changed to a 8-in. block wall, and a 10-in. block wall would be changed to a 6-in. block wall. The inside of the foundation remains even while the outside forms a 4-in. shelf.

Using Solid Block on the Last Course

In some areas, building codes require that the last course of block be solid or that the holes be filled in the top of the block to help distribute the load. I recommend using solid block rather than filling up the cells with mortar and scraps of block. Anchor bolts to hold the wood sills down are installed at the proper intervals between the head joints of the block and are cemented in place. The bolts should be either 7 in. or 15 in., depending on local building codes, and have an L-bend on the bottom so they do not pull out. The wood plate is usually 1½ in. thick, so I recommend letting the bolts extend about 2 in. above the top of the block wall to allow for tightening the nuts. The use of rebar or horizontal rebar reinforcement is usually only required in areas prone to earthquakes, hurricanes, or high winds.

Waterproofing the Foundation

The traditional method of waterproofing a concrete block foundation is parging (plastering) two separate coats of a portland cement/lime waterproof mortar on the block and then, after it has cured, applying an asphalt emulsion compound over that. Both of these are only part of a total waterproofing system used by builders. If combined with proper grading away of the earth from the foundation, a good footing drainage system, and properly positioned drain spouts that don't empty near the walls, the chances of having a dry basement are excellent.

There is a variety of mortar mixes that masons use for parging foundations. The parging mortar mix that I prefer and recommend is a portland cement/lime mix in the proportion of 1 part portland cement to ¼ part hydrated lime to 4½ parts sand. This is stronger than regular mortar and is classified as Type M. It is recommended to be used below grade. Make sure you ask for waterproof portland cement at the supplier rather than regular portland. It is formulated especially for waterproof mortar.

The cleaner the wall is, the more effective the parging will be. Before starting any parging, clean around the bottom of the walls using a stiff broom or brush. Chip

Using a plastering trowel and aluminum hawk makes parging a wall go quickly.

away any hard droppings of mortar with the brick hammer. Scrape any protruding mortar from the walls so they do not interfere with applying the parging.

I recommend using a plastering trowel rather than a regular bricklayer's trowel to parge the mortar on the wall because it will do a better job and there is a heck of a lot less strain on the wrist. Plaster trowels are rectangular, like a cement-finishing trowel but a little shorter in length. You can either make a plaster hawk out of plywood or pick one up at a building supply center. The mortar is picked up from the wheelbarrow or mortarboard, placed on the hawk, and then plastered on the wall with the trowel from the hawk, which saves constant trips to the mortarboard. It works great because you can hold the hawk next to the wall and catch mortar dropping from the trowel before it hits the ground.

When parging a wall, start by forming a mortar cove at the bottom of the wall.

Trowel the parging mortar on the wall, using long steady strokes to keep it smooth.

Parging the walls Dampen the block walls first with a fine mist of water from a garden hose with a nozzle on it so that the parging does not dry out too fast and crack. Don't soak the walls! Start at the bottom of the wall and trowel a thicker cove, or angled coat of mortar, at that point to help drain water away from the bottom of the wall. Next, trowel the mortar up the wall with long strokes, keeping it as smooth as possible and approximately ¼ in. to ⅜ in. thick.

After the parging has been troweled on and before it dries, scratch the surface with an old straw broom or a scarifier, which is a special tool made for this purpose that resembles a large metal comb. This is done to ensure that the second coat will bond to the first coat.

Let the parging dry overnight, then wet it with a fine spray of water, and trowel on the second coat of parging, making it as smooth as possible. Even out any rough spots by dampening a medium soft brush and brushing the parging lightly.

Scratch the first coat of parging with a scarifier before the parging dries.

Smooth out the final coat of parging with the trowel.

Applying asphalt emulsion waterproofing compound Complete the waterproofing process by applying two coats of an asphalt emulsion waterproofing compound over the parging. It can be applied with either a stiff brush or sprayed on with high-pressure equipment. A product I recommend highly that has been used by builders in my area for many years with excellent results is called Hydrocide 700B (Sonneborn Building Products, 57-46 Flushing Avenue, Maspeth, New York 11378). It comes in 5-gallon buckets and is available at most building suppliers. I prefer it because it stays a little tacky, resists cracking, adheres to the block wall very well, and has a proven track record. As you can imagine, applying any type of tar or asphalt base waterproofing is a messy job.

The wall should be dampened with water before applying the compound. Let the first coat dry to a tacky state before applying the second coat. Use a long-handled brush and wear rubber gloves and old clothing, when applying it. Hydrocide 700B can be cleaned from tools or the sprayer with water while it is still fresh. Be sure to read the directions thoroughly on the can before using it.

A completed and waterproofed concrete block basement foundation.

Section View of a Typical Concrete Block Basement Foundation Wall and Footing

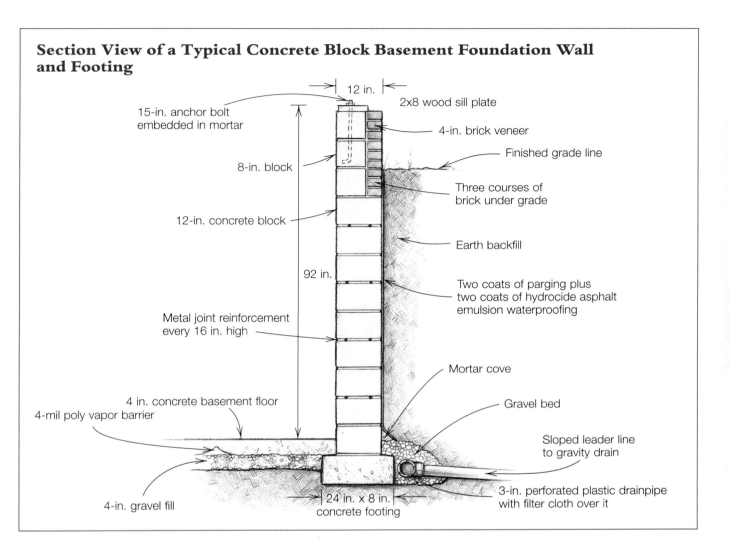

12 in.

2x8 wood sill plate

15-in. anchor bolt embedded in mortar

4-in. brick veneer

Finished grade line

8-in. block

Three courses of brick under grade

12-in. concrete block

Earth backfill

92 in.

Two coats of parging plus two coats of hydrocide asphalt emulsion waterproofing

Metal joint reinforcement every 16 in. high

Mortar cove

4 in. concrete basement floor

Gravel bed

4-mil poly vapor barrier

Sloped leader line to gravity drain

24 in. x 8 in. concrete footing

3-in. perforated plastic drainpipe with filter cloth over it

4-in. gravel fill

Placing drainpipes and drainage materials All of the good work you have done will be to no avail unless water is drained away from the exterior of the foundation walls! Almost all building codes that I know require some type of drainage materials around the foundations and footings to prevent water from pooling there. There are different methods of accomplishing this. The most effective one I have used is to spread a layer of crushed stone or gravel around the foundation next to the footings and then lay 3-in. flexible perforated drainpipe on top of it. Covering the top of the pipe with a filter cloth prevents mud and earth from blocking the holes. Another layer of crushed stone is placed on top of that.

If drainpipes are to be effective, the water that collects in the pipe should be drained to a gravity drain installed on a slope a distance away from the foundation. The worst thing you can do is merely trap the water in the drainpipe and leave it to collect around the foundation. Different areas of the country have unique drainage problems, so before installing any footing or foundation drains, always check the local building codes for their specific recommendations. In my opinion, you can't overdo the waterproofing process to obtain a dry basement. In many areas of the country, builders of homes must guarantee a dry basement for one year.

Chapter 7

FUNDAMENTALS OF CONCRETE WORK

Concrete plays an important part in almost all masonry projects. It not only supports tremendous loads but it can also be poured in almost any shape or form. Concrete outlasts any other building material I know of and requires no maintenance or upkeep. Dollar for dollar spent, it probably is the greatest bargain in construction work. It's hard to imagine a masonry project that does not utilize concrete somewhere.

As you read through this chapter, keep in mind that once concrete is in place, you are committed! The only way it can be torn up or changed is with a jackhammer. This is not only time-consuming but also very costly.

Concrete work can become a very technical subject in a hurry. My purpose in this chapter is to cut through the technicalities and cover a variety of useful background information. I explain how to mix concrete, how to estimate concrete needed for a job, how to prepare to pour concrete, factors to consider when placing your order from the plant, tools and equipment you need, proper curing procedures, and how to avoid problems.

COMPOSITION OF CONCRETE

Before starting to work with concrete, it is helpful to have a basic understanding of the materials that make it up. Concrete is made basically of a mixture of portland cement, fine and coarse aggregates, and water. The

Being prepared to spread the concrete as soon as it arrives saves time—and money.

active ingredients are the portland cement and water. When mixed together, they start a chemical process known as *hydration,* in which they change from a thick liquid mixture into a dense solid mass after curing. After concrete has been poured and finished for a day, it may appear to be dried out. The truth is only the surface area has set. It takes 28 days for concrete to cure to a testing strength.

Portland Cement

Portland cement is not a brand name but a type or grade of cement. Portland cement was invented in 1824 by Joseph Aspdin, an English mason. He made the first batch of portland cement by heating a mixture of clay and limestone in an oven and grinding the resulting clinker. He discovered that when mixing this with sand, stones, and water, it formed a material that became extremely hard, even under water. The name portland was given to the cement because it was gray and resembled building stone quarried near Portland, England. Portland cement is produced today in huge rotary kilns, basically following the original procedures.

Don't be confused when buying portland cement! Although there are five different types of portland cement made for mortars and concrete, I would stick with Type 1, which is a general all-purpose cement and is the most commonly used. The rest are used for design mixes to achieve special results. Unless you request another type, the salesperson at the masonry supplier is going to sell you Type 1 anyway.

Portland cement is packaged in 94-pound bags, which contain enough mix to make 1 cu. ft. of poured concrete. For many years, portland cement was sold by the barrel, which contained the equivalent of four bags. Old habits are hard to break, and many masonry suppliers still write barrels on the sales slip when selling more than four bags. It is priced, however, by the bag regardless of what is printed on the sales slip.

The primary function of portland cement in concrete is that when mixed with water, it forms a cement paste that fills the voids between the fine and coarse aggregates and becomes one solid hard mass. It is important to know that the less water added to the mix, the stronger the concrete will be, and the more water added, the weaker the concrete will be. This is known as the *water/cement ratio,* which is usually stated in gallons of water per 94-pound bag.

The general rule I follow when mixing concrete is to add as small an amount of water to the mix as possible that allows the materials to blend and still permits me to fin-

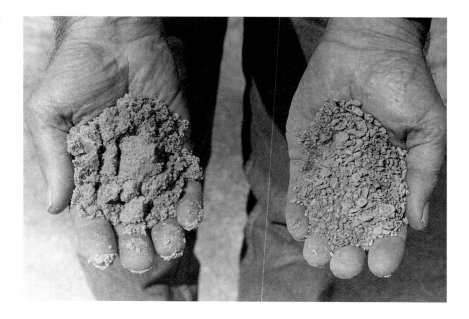

An example of fine sand used for mortar is shown in the left hand and a sample of fine stone screening aggregates used for concrete is shown in the right hand.

ish it as desired. As a rule you can figure that on the average a footing-grade concrete takes about 6 gallons of water per bag in the mix, and a finished floor or slab takes about 5 gallons per bag. Concrete should have a creamy gray texture whether it's in a footing or a floor mix.

Aggregates

The fine and coarse aggregates in concrete should be reasonably clean and free of any type of soil, such as loam or silt. The fine aggregates, either sand or fine crushed stone screenings, can range in size from fine up to ¼ in. Although regular building sand can be used for concrete, a silica or coarser-grain sand is better. Masonry materials suppliers I deal with keep mortar and concrete sands in separate bins so there is no mixing them up.

Coarse aggregates can be either gravel or crushed stone, ranging in size from ¼ in. up to 1½ in. In my immediate area, we are blessed with an abundance of natural blue limestone, which is one of the finest aggregates for concrete there is. Crushed stone or gravel that is angular or rectangular in shape is better than rounded or oval gravel because the cement paste will adhere to it a lot better. In some areas of the country, footing concrete mixes use

a 1½-in. stone (because footings are not troweled or finished) and a ¾-in. crushed stone aggregate for finished concrete. The local concrete plant that I deal with uses ¾-in. crushed limestone for both footing and finish concrete for residential applications.

Water

The general rule for water to mix with concrete is that if it is clean enough to drink, then it is okay to use. Regular tap water is fine. If you live near saltwater, be careful about using it for concrete, as the salt can affect the ultimate strength of the mix and may corrode any steel reinforcement in the concrete. The same is true for beach sand, which has a lot of salt in it.

Admixtures

Admixtures are any materials added to the concrete during the mixing process other than portland cement, aggregates, and water. There are four major admixtures that are used frequently and give concrete special properties.

The first are known as *air-entraining agents* and are popular in areas of the country that experience very cold weather. Air-entrained concrete is made by using air-entrained portland cement or by adding an air-

The coarse crushed blue limestone aggregate shown here ranges in size from ¾ in. to 1½ in.

entraining agent to the mix at the plant. This additive traps and stabilizes billions of air bubbles in the concrete creating small voids or cushions, which in turn permit the concrete to resist freeze-thaw cycles, acting like miniature shock absorbers on a car. Concrete with air-entraining agents has a creamier texture than regular concrete and has improved workability, which is a big asset in the finishing process. The additives also cause the concrete to be more resistant to salts and cracking. It is highly recommended for sidewalks and slabs that are subject to cold weather.

There are also special retarding agents that can slow the setting time by 30 percent to 60 percent. They are used when pouring concrete on very hot days and when you don't want the concrete to dry too fast. These are usually added by concrete specialists at the plant as special orders.

Water-reducing agents can also be added. Known as *plasticizers,* these allow as much as 15 percent reduction in water volume to achieve a specified slump requirement. (The mix consistency or degree of stiffness of green, or uncured, concrete before it cures is called its *slump.*) Plasticizers therefore tend to increase the strength of concrete and create a better bond to steel reinforcement rods.

Lastly, the well-known additive *calcium chloride* (commonly called *Cal* in the trade) speeds up the set of concrete in cold weather. Ideally, the best curing temperature for concrete is about 70°F, but in many cases this is not possible. If concrete is permitted to freeze when it's setting or curing, it will lose a lot of its specified strength and a host of problems can result, such as cracking, scaling, or dusting. Calcium chloride should only be used if you cannot avoid pouring concrete in freezing weather. When ordering concrete with calcium chloride added, request that it be no more than 2 percent by weight of concrete. This material, if used in excess of recommended amounts, can cause the corrosion of any reinforcing steel in the project.

ESTIMATING CONCRETE

Concrete is estimated and ordered in cubic yards. A cubic yard contains 27 cu. ft. For example, to estimate the amount of concrete needed for a footing, first find the total number of linear feet around the walls of the foundation, including features such as porches or piers. Multiply the length of the footings in ft. x the thickness in ft. x the width in ft. This will give you the number of cu. ft. in the footing. Divide this figure by 27 to find the number of cu. yd.

The formula is expressed as follows:

$$\frac{\text{Length in ft. x thickness in ft. x width in ft.}}{27} = \text{cu. yd.}$$

You may find it easier to express inches as fractions or as decimals of a foot in your calculations. (See the chart on p. 136.)

Expressing Inches as Fractions and Decimals of a Foot

Inches	Fractional Part of a Foot	Decimal Part of a Foot
4	⅓	0.33
5	⁵⁄₁₂	0.42
6	½	0.50
7	⁷⁄₁₂	0.58
8	⅔	0.67
10	⅚	0.83
12	1	1.00

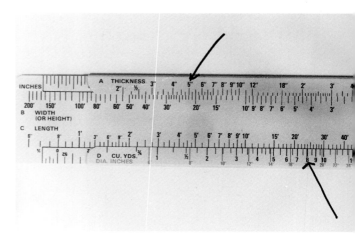

This concrete slide rule is set up to show the calculations of the example given in the text.

The worst thing you can do when estimating concrete is to come up short! If the footings are not excavated square or are erratic in width, take this into account. My suggestion is to check each one individually by getting down into the trenches and measuring. Mark down any sizeable differences on a notepad, and take them into account when coming up with the final figure.

There is an old saying in the concrete business: "If you do not have a little extra, you do not have enough." Running short means that you have to mix some yourself, and that is a big hassle when you want to get the job done. Generally allow 5 percent more than you estimate to fill in any places that sink or are not true to size. Prepare an extra place in advance to utilize any leftovers, such as a short length of sidewalk, a base for the garbage cans, or a place to sit the picnic table on. The truck driver will appreciate this, and you will receive full value for your concrete dollar.

Estimating with a Concrete Slide Rule

Rather than get into a lot of involved math, there is a simple method of estimating concrete using a concrete volume slide rule. Almost all concrete companies give them away free for the asking as advertising. The instructions for using them are simple and are printed with an example on the center section of the rule.

To illustrate how a concrete slide rule works, I'll use the example of a typical concrete garage floor that measures 5 in. thick x 22 ft. wide x 24 ft. long. The thickness of the concrete floor on the slide rule is represented by A. Slide the center section B (width) until 22 ft. lines up with 5 in. Now look at section C (length) and you see that 24 ft. lines up with 8 cu. yd. on section D (total cubic yards). (See the photo on this page.) Adding 5 percent for low spots (an extra 0.4 of a yd.), round this off to an extra half a yd., making the final figure 8½ cu. yd.

ORDERING CONCRETE FROM THE PLANT

There are a number of important factors to consider when placing an order for concrete from the plant. The most important is what type of mix you want. Although you can order a number of various mixes with different strengths, there are two general formulas or ratios for concrete that should serve all your needs. For footings, I use a 1:2:4 mix, which means 1 part portland cement to 2 parts fine aggregates (sand or fine crushed stone screenings) to 4 parts crushed stone or gravel with about 6 gallons of water per bag of portland cement. This is considered to be a five-bag-to-the-cubic-yard mix.

For concrete slabs, floors, or flatwork, where you want more cement paste on top for a good smooth troweled finish, I recommend a 1:2:3 mix of 1 part portland cement to 2 parts fine aggregates to 3 parts crushed stone or gravel with about 5 gallons of water per bag of portland cement. This is considered to be a six-bag-to-the-cubic-yard mix.

The other method of ordering is commonly called the *prescription mix*. This is where you specify the psi (pounds per square inch) strength of the cured concrete that you want. If the dispatcher uses this method, keep in mind that a five-bag footing mix equals 2,500 psi, and a finish concrete mix of six bags per cubic yard equals 3,500 psi.

Both methods of ordering concrete give the same results. As stated before, all concrete strengths are based on a 28-day curing cycle.

It is a good practice to place your order at least two days in advance. By all means avoid ordering concrete after regular working hours or on Saturdays when there will surely be a hefty extra charge. Concrete companies normally allow from 30 minutes to 60 minutes to empty a full truck (8 cu. yds.) before charging extra. Ask how much unloading time is allowed when you place your order.

Concrete trucks come equipped with standard 10-ft. or 12-ft. metal chute sections to unload. If that will not reach far enough, inquire if they have a truck that will pump concrete or run it on a long belt to the forms. Some companies have these, but there is an extra charge for them.

If you need less than a full truckload of concrete, ask if there is a minimum short-load hauling charge. In my area, loads less than 5 cu. yd. are considered to be short loads. I would also inquire if there is a mileage charge from the plant to the job site. Most companies have fixed mileage zone rates from the plant included in their standard price. Mileage beyond that is charged to the customer.

Be sure to check out the weather report before placing your order. If it is calling for rain, you may want to move delivery up a day. When pouring footings or concrete that does not have to be finished, the weather is not as critical, but it can be a major headache to try to trowel a floor or sidewalk smooth in the rain.

PREPARING FOR THE CONCRETE TRUCK

You can cut down a lot on frustration and last minute headaches by having the job ready to pour. It is impossible to control all of the variables but preplanning can take care of most of them. The first one is to have plenty of help available as you never know what may crop

Use pliers to cut wire that may stick up through the concrete.

up. Keep in mind that the driver will only unload the concrete through the chute. His responsibility ends there.

Have all of the tools and equipment handy that you need to move, pour, and finish the concrete, along with a roll of plastic covering in case it rains. You will need a garden hose with a nozzle close by to rinse off the tools as soon you are finished with them. Rubber or gum boots are a must because at some point, you may have to wade into the concrete to float or screed it.

For floors, slabs, or sidewalks, have the necessary wood forms in place and properly braced. Lay a roll of plastic vapor barrier (4 mil) on the base to prevent water in the concrete from being absorbed into the ground prematurely and not allowing the concrete to cure properly. Place a roll of concrete reinforcement wire on top of the plastic. I always keep a pair of heavy-duty cutting pliers and bolt cutters handy in the event that a piece of wire sticks up stubbornly through the concrete (as shown in the photo on p. 137). If you plan to wheelbarrow any concrete, establish the best route in advance. It may require laying some boards down so the wheelbarrow does not sink into the base.

Plan the path the concrete truck is going to follow to get into your place without causing damage or getting stuck. A truck fully loaded with 8 cu. yd. of concrete weighs about 30 tons. You cannot run it across water lines, septic tanks, or drain fields without expecting some damage.

MIXING YOUR OWN CONCRETE

Relatively small amounts of concrete can be mixed on the job in a wheelbarrow or a utility mixer. By small, I am talking about ½ cu. yd. or less. I have always found it astonishing that 1 cu. yd. of concrete weighs approximately 2 tons. It is a lot of hard work to mix, even with the aid of a utility mixer, and out of the question as far as I am concerned with a mortar hoe.

Start by setting up the mixer level, close to the sand and crushed stone pile. I always pick a shady spot, if possible, away from the house where spilled concrete will not ruin the yard or grass. The lady of the house will appreciate this! Hook up a garden hose with a nozzle, and gather together a few 5-gallon buckets.

Tools and Equipment You Will Need

You only need a relatively small number of tools and basic equipment to handle and pour concrete. There may be a few that you need to buy or borrow, but they are not that expensive and are stocked by most building suppliers.

Metal wheelbarrow
(5 cu. ft. size)

Straight wood screed board (2x4 or 2x6)

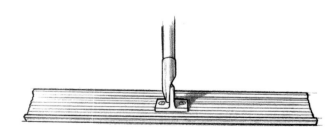

Metal bull float with extension handles (can be rented)

Bricklayer's trowel
(can be an old one)

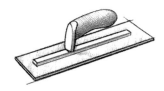

Steel concrete finishing trowel

Metal float (magnesium works best)

Wood float (excellent for floating a rough
spot to bring cement paste to the top)

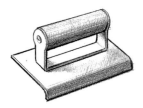

Edger (used to round off
edges of slab or sidewalks)

Cement groover (used to form a
groove at expansion or control joints)

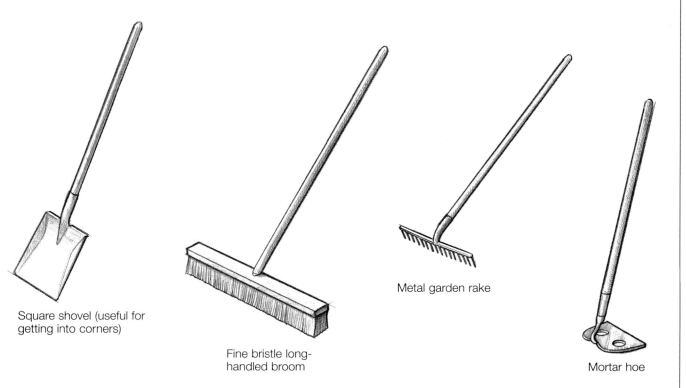

Square shovel (useful for
getting into corners)

Fine bristle long-
handled broom

Metal garden rake

Mortar hoe

One of the problems with utility mixers is that the materials stick to the sides. The best way to avoid this is to pour in 2 gallons of water first, followed by all of the larger crushed stone. Let the mixer turn a few revolutions and then add the sand or fine aggregates and the portland cement slowly, adding more water as needed. Let the mixer run for four or five minutes until the ingredients blend together into a gray uniform mixture. Adding water to the mix from a garden hose is not very accurate. I prefer to measure it from a 5-gallon bucket, as I can keep track of the amount better.

For most small utility mixers, a good size batch would be three shovelfuls of portland cement, six shovelfuls of sand or fine aggregates, nine shovelfuls of gravel or crushed stone along with about 2 gallons of water. (The amount of water needed varies according to the moisture content of the sand.) If the mix does not look rich or creamy enough, it never hurts to add an extra shovelful of cement. You will learn in a hurry never to look into the mixer drum while it is turning—this could be compared to looking into the mouth of a cannon ready to fire! Concrete will invariably slop out into your eyes or face or onto your clothes.

When you are finished mixing the last batch, throw a couple shovelfuls of crushed stone in the mixer drum, and spray some water from the garden hose in while it is still running to scour it out well. Then empty this out, and rinse the mixing drum with the garden hose.

POURING FOOTINGS

There are two types of concrete footings—trench and formed. In a trench footing, the sides of the excavation act as natural forms holding the concrete in place. This is by far the simplest method to install. If you are going to excavate anything larger than a garden shed, I strongly recommend getting a backhoe on the job. The proper size of foundation footings was briefly discussed in Chapter 6. Remember, average footings are about 8 in. deep and twice as wide as the block width to be used.

Depending on the width of the trench, the backhoe operator can use different-size buckets to match the footing width you want to pour. The more vertically true the trench is, the stronger the poured footing will be, as it will distribute the load more evenly. The bottom of the trench should also be reasonably level.

Before pouring any concrete into the trenches, determine the finished height of the footing, and drive short lengths of steel reinforcement rods level with each other in the center of the trench approximately 5 ft. apart. These serve as leveling points for the concrete. I usually

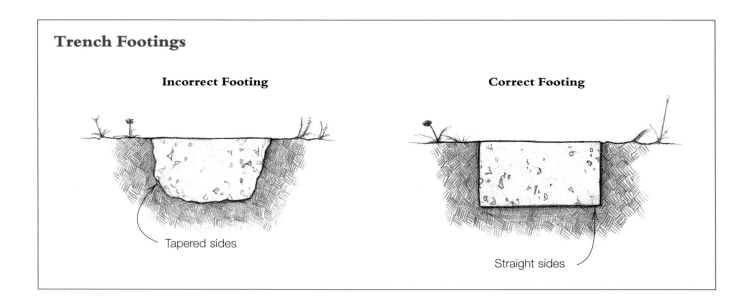

Trench Footings

Incorrect Footing

Tapered sides

Correct Footing

Straight sides

Wood-Braced Form for a Concrete Footing

Note: A piece of ¾-in. plywood can be used in place of 2x8 framing lumber if braced properly.

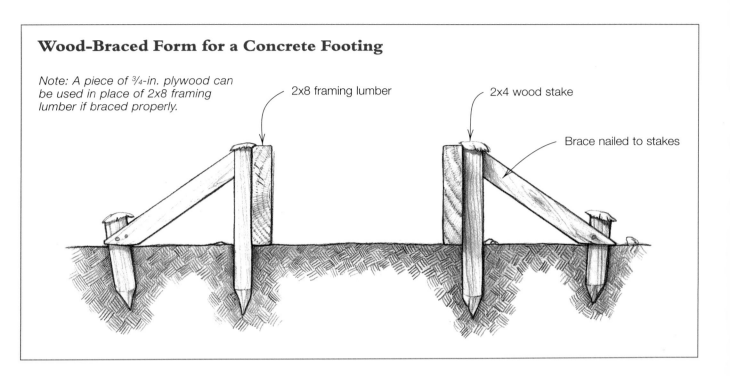

2x8 framing lumber

2x4 wood stake

Brace nailed to stakes

Using Steel Rod Stakes as Guides for Concrete Height

A short steel rod driven in the trench serves as a leveling point for concrete.

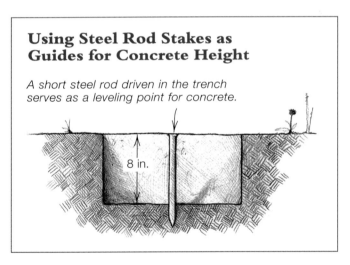

8 in.

Unless there is some special problem, horizontal steel reinforcement rods are not usually installed in footings. The top of the concrete should not be smoothed out but merely leveled with the end of a rake. This establishes a better bond with the mortar joint when the block are laid.

A formed footing is required when rock or very hard ground is encountered. On large construction projects, steel forms are used. On small jobs, framing lumber works fine. The lumber should be thick enough and braced sufficiently to prevent it from bowing as the concrete is poured in it. I recommend using 2x8 framing lumber for forms if available. The trick is to make sure that the forms are braced well.

Sometimes when rock is encountered and you can't excavate much, the top edge of a formed footing will show above the basement floor inside the foundation after it's completed. You can keep this smooth and free of voids by tapping against the edge of the form with a hammer as the concrete is poured. This will help prevent what is

drive stakes in their proper positions on each corner of the foundation, stretch a nylon line tightly between them even with the top of the steel rod, and drive the rest of the stakes in place even with the line. Once the concrete is poured in the trench, the steel rods can be left in place. Wood stakes are a poor choice as leveling points as they will rot.

Typical trench concrete footing for a house foundation.

commonly called *honeycombing*. After the concrete has set overnight, the forms can be removed and cleaned so they can be used again.

FORMING AND POURING A CONCRETE SIDEWALK

A sidewalk is a typical example of concrete flatwork. Whether it's a driveway, basement floor, or sidewalk, the procedures and techniques of pouring and finishing concrete are basically the same. My brother-in-law decided to build a swimming pool in his backyard with connecting concrete sidewalks. Since most of the family work in construction trades, we all decided to pitch in and do the work in exchange for swimming rights. It turned out to be a great deal for all of us.

Preparing the Forms

We started by building and setting the wood forms. The average depth of a concrete sidewalk is about 4 in. Wood 2x4s work fine for sidewalk forms; however, keep in mind that they are actually 3½ in. You want to excavate a tad deeper to compensate for this. We installed the 2x4 forms by driving wood stakes in a line, even with the top of the existing grade. We then nailed the wood forms even with the top of the stakes. I prefer to use double-headed concrete nails, as they are easier to remove from the forms after the concrete has set. Allow a little pitch or slope (known as a *fall*), about ¼ in. per ft. for small areas like sidewalks, to permit water to drain off. (For larger areas, ⅛ in. per foot should be used.) We did not have to allow any fall because the sidewalk was on a descending grade.

Long concrete sidewalks are notorious for cracking due to expansion or settlement, so we placed temporary 2x4 dividers at 10-ft. intervals across the walk to allow for some movement. We set a piece of 30-pound roofing felt against the divider to act as a control joint. As the concrete was poured, we removed the dividers but left the roofing felt in place.

Typical Concrete Sidewalk Forms

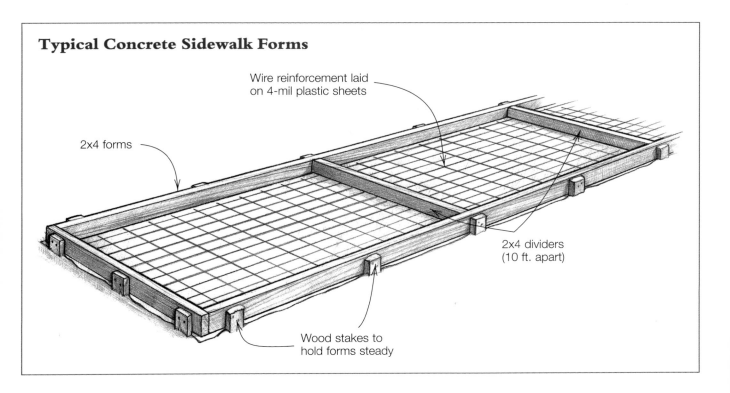

Wire reinforcement laid
on 4-mil plastic sheets

2x4 forms

2x4 dividers
(10 ft. apart)

Wood stakes to
hold forms steady

The forms for a concrete sidewalk are in place, complete with plastic underlayment and reinforcement wire.

If the soil beneath the walk is firm and undisturbed, you only need to lay a plastic vapor barrier on it to prevent the water in the concrete from being absorbed too fast by the dry earth. I recommend using 4-mil plastic, available in rolls from any building supplier. I like using plastic as an underlayment because it lets the concrete set and dry out evenly, making it easier to finish. We completed the form preparation by laying concrete reinforcement wire on top of the plastic to prevent the concrete from cracking. If you want to curve the forms here and there, use strips of ½-in. plywood. Just be sure to brace it well so it does not move when the concrete is poured against it.

Pouring and Finishing the Concrete

When the concrete truck arrives on the job, locate it as close to where you want to pour as possible. In this particular case, we could not get right up to the forms, so the truck was parked in the driveway and wheelbarrows were used to transport the concrete to the forms (see the left photo on p. 144).

The concrete truck has its chute in place ready to discharge concrete.

The concrete is worked into the forms with a shovel straight from the wheelbarrow.

We started at one end of the walk and dumped the concrete into the forms, being careful not to rest the wheelbarrow against them. It should be worked well into the corners and along the edges with a shovel and rake, leaving it just a little high to screed off. A bricklayer's trowel comes in handy for working along the edges. The technique that I use is to insert the edge of the trowel blade between the edge of the form and the concrete and work it up and down in a slicing motion. Tapping on the edge of the forms with a hammer also helps to fill any voids and make a smoother edge.

It is important not to dump the concrete in piles but always right against what was previously poured so there are no places to fill in. You need one or two

people working the shovel and rake as it is being poured from the wheelbarrow (see the right photo above). As the concrete is being poured in the forms, the reinforcement wire should be raised up a little with the end of the rake, so it will be well embedded throughout the concrete.

Screeding the Concrete

As the concrete is being poured in the forms, another person should be constantly screeding it off even with the top of the forms with a 2x4 (see the photos on the facing page). The wood screed should be a little longer than the width of the walk so it does not slide off the edge. For large areas such as a floor, longer straightedges should be made of 2x6 lumber to prevent them from

Use a 2x4 to screed the concrete on a sidewalk.

Use a longer straightedge board to screed wider sections.

riding up on top of the concrete too much. Screeding is done by moving the screed back and forth in a sawing action, letting it rests on the top of the forms at all times. Any low spots are filled in as the concrete is screeded. Excess concrete that builds up in front of the screed is pulled forward with a rake. On a walk, only one person is needed to do the screeding.

In addition to making the concrete level with the forms, the screeding forces all of the aggregates down into the mix and brings the cement paste up to the top. This paste is known as the *cream* and has to be on the surface to help fill all the voids. It can be finished later. The more cement paste that is on top, the easier the concrete will be to trowel smooth.

As we worked our way to the 2x4 divider strips, we gently removed them and filled in the voids, leaving the strip of tar paper in place for the control joint (see the photo above).

Floating the Surface
As soon as you see the standing surface water start to disappear and the concrete lose its sheen, the concrete should be floated. Floating is done to work up more cement paste to the surface and make sure that all of the aggregates are completely imbedded in the mix. How long this takes depends on the prevailing temperature and weather conditions.

Floating is performed by moving the float in a circular troweling action against the surface until the cement paste works up on top of the surface and creams over (see the top photo on the facing page). I prefer a metal magnesium float as it is lighter and smoother. I save the wood float for stubborn spots, where the paste does not come up readily, as wood tends to pull it up better. As soon as you have finished floating, rinse the tools with water, so they are ready to be used again.

For areas that you can reach across, the hand float works fine, but for larger areas, a metal bullfloat tool that has extension handles is a necessity (see the bottom photo on the facing page). The bullfloat is pushed across the surface on a slight angle so it does not dig in and then is drawn back with a little hopping action of the wrist. It only takes a couple of minutes to get the hang of it, and it is the only way to go when floating a large slab or floor.

Floating the concrete is done to bring cement paste to the top for troweling.

Edging and Grooving

As soon as the floating is complete and there is plenty of cement paste present on the surface, all open edges should be rounded off with an edging tool to seal and prevent them from chipping. Run the edger along all of the edges with a nice steady movement. Try not to let the edger dig into the concrete; use more of a gliding action. You don't want to form too deep an impression, just a rounded smooth edge (see the photo on p. 148). After trying it a few times, you will discover that if the front edge of the tool is held up just a little, digging in will be held to a minimum. I find that after the concrete sets a little more and some of the water evaporates, edging a second time does an even better job.

Grooving is done the same way as edging, only it's done along the edges of the control joints where the tar paper is located. Its purpose, as the name implies, is to cut or form a smooth groove, approximately ⅜ in. to ½ in. deep, that will prevent the edges from chipping, or *spalling off*. To form a nice straight groove, lay a straight guide strip of wood along the joint across the walk. Let the groover ride against the edge of the strip and run it

Using the bullfloat is necessary for floating large areas of concrete.

Round the edges with an edging tool to seal them and to keep them from chipping.

along the edge of the joint, pressing down hard enough to cut the groove. After floating and grooving, refloat the entire surface area to smooth out any impressions, being careful not to interfere with the rounded edges.

Finishing the Surface

There are two basic ways to finish a concrete surface—smooth, or coarse and gritty. Most concrete floors, sidewalks, and slabs are troweled smooth as they are easier to keep clean that way. The trick is to know when to start troweling to obtain a smooth finish. The first troweling should be started when the water sheen starts to disappear from the surface. The only way you can tell when that occurs is to stay there and watch it.

I test to see if the concrete is ready by touching the surface with the end of my finger, and if it leaves a slight impression, it is time to start troweling. A cardinal rule to follow is never leave the concrete when you are waiting for the right time to trowel it. Timing means everything when finish-troweling concrete, as the cement paste has to be smoothed out at just the right time. The only way I know to learn this is by trial and error. As the old expression goes, "A fast train waits for no one."

Forming a Groove Joint between Two Sections of the Walk

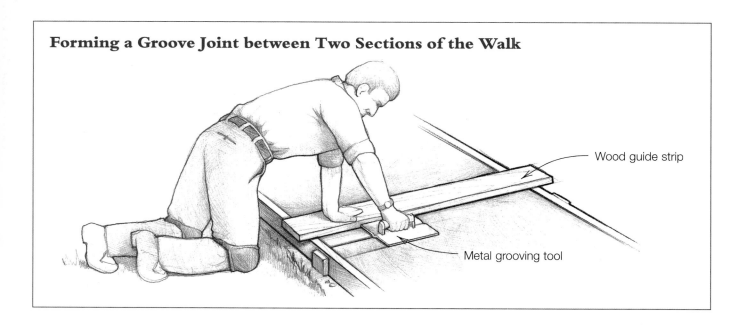

Wood guide strip

Metal grooving tool

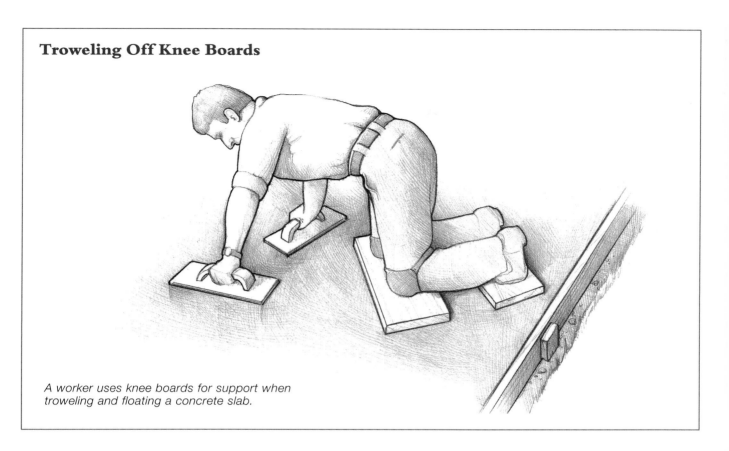

Troweling Off Knee Boards

A worker uses knee boards for support when troweling and floating a concrete slab.

Trowel with long sweeping strokes, holding the blade up on a slight angle. The idea is to smooth out the surface as much as possible without digging in or leaving a lot of trowel marks. There should be a rich creamy texture of cement paste left on the surface at the completion of the first troweling. If not, refloat any areas that need it.

If the area is too large to reach across, wood knee boards (also called kneeling boards) work well. A couple of pieces of ¾-in. plywood about 12 in. x 24 in. work fine. One board is for your knees and the other for your feet. Raised 1-in. strips of wood nailed along one edge enable you to pick them up and move them as needed. They should support your weight without sinking in. I generally work my way out to the center of a floor on knee boards and trowel my way back to the edge, smoothing out any impressions the boards left as I retreat.

Finish-trowel the concrete by using long sweeping strokes and holding the trowel blade up slightly.

Creating a Nonslip Finish

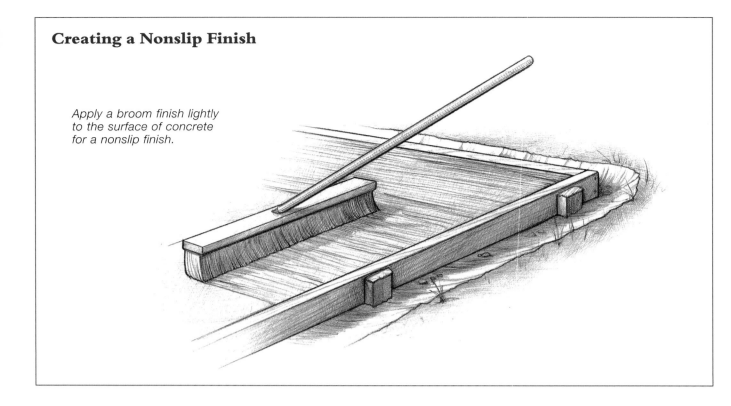

Apply a broom finish lightly to the surface of concrete for a nonslip finish.

After the surface concrete has dried off enough from the first troweling, it can be retroweled to smooth out any areas that need it. How many times you retrowel for a smooth finish is a matter of personal choice. As a rule, it takes me about three trowelings to obtain the finish I like. Although I have described how to finish concrete here, only doing it yourself will really teach you how.

Brushed finish Sometimes a nonslip surface is desired on sidewalk or driveway. I don't particularly like this kind of finish because it collects too much dirt. However, this can be achieved by lightly brushing the surface with a fine brush after the first troweling dries a little. A long-handled floor broom works best. The brushing can be wavy or straight, whichever you like. If the surface is dry, flick a little water on it with a brush before using the broom. You want the finish to be only gritty, not rough.

Curing the Concrete

If concrete is to reach its full strength, it needs to cure or season properly. Assuming that it was mixed correctly, the temperature and weather conditions are the most important factors to consider. Don't be fooled by outward appearances—although it may look like it is dry, finished concrete should not have any traffic on it until at least three days have passed and probably a week for driveways.

In warm weather, spray a mist of water on it a couple times a day, so it won't dry out too fast. If you are going to be away for a couple of days, wet it and spread a plastic vapor barrier on it to hold the moisture in. In cold weather, spreading straw on top helps as the straw tends to insulate and cut down on heat lost from the hydration process. If you have access to burlap, lay it on top and wet it well with water. The most important thing is for the concrete to cure slowly and not dry out too fast.

Three Major Problems in Concrete

There are three main problems that can develop in concrete, which you want to avoid. The first is called *scaling*. Scaling occurs when the top or surface shreds or flakes off to a depth of 1/16 in. to 1/4 in. Most of the time this occurs because of freezing and thawing. Scaling can also be caused by letting the concrete dry too long before finish troweling and then sprinkling too much water on the surface when it is being troweled.

The second one is known as *crazing*. This is defined as many fine hairline cracks on the new concrete surface after it has set. The cracks look a lot like crushed eggshells. This condition is due primarily to excess shrinkage, which can be caused by the surface drying too fast under a very hot sun or in windy conditions. Finish-troweling concrete too soon when there is a lot of surface water present also contributes to this problem. What happens is that this tends to float the fine aggregates such as sand to the surface where they dry.

The third is *dusting*. This is probably the most common condition. Dusting occurs when a dry powder collects on the surface. Letting the concrete set too long before finish troweling is in my opinion the number-one reason for this condition. A half-hearted corrective measure of sprinkling dry portland cement on top to work up additional paste and then troweling it in also worsens the problem. Almost all of these problems I have discussed here can be avoided by following the proper finishing procedures at the right time.

The completed sidewalk leading to the swimming pool.

Chapter 8

MASONRY REPAIR AND RESTORATION

While it is true that a masonry building requires less up-keep than other types of structures, as the years pass, it is inevitable that some type of repairs will be needed to maintain its good condition, protect your investment, and prevent little problems from turning into big ones. The possibility of a wall cracking or leaking in a new home is rather remote, but in older homes, repairs for such problems are commonplace. Over the years, I have diagnosed and repaired a large variety of masonry problems. Most of the questions and inquiries that I receive from the readers of trade magazines that I write for concern repairs rather than new construction. The most frequent masonry repairs are cutting out and repointing mortar joints in older buildings, repairing cracks in masonry walls, removing paint and stains from masonry walls, treating leaky basements, waterproofing the exterior of brick walls that leak from rain, replacing retaining walls that have pushed out or fallen over, and diagnosing and repairing faulty chimneys and fireplaces.

In this chapter I share with you some of the more common problems that I have come in contact with, point out possible causes, and explain procedures, techniques, and tips for repairing them. Most often, a visual inspection provides the best clues to identifying the cause of the problem.

CUTTING OUT AND REPOINTING MORTAR JOINTS

Old brick buildings don't have to just fade away and die. They can be restored and made as good as new by repointing with fresh mortar. Most brick buildings that

Repointing mortar joints in a brick wall can renew a deteriorating building.

This brick wall is in poor condition and needs repair and repointing.

need repointing generally date from the period when lime and sand mortar was used, which contained very little or no portland cement. The brick were softer than those made today, as many were handmade in molds and not fired in kilns. Many old brick are beyond saving and will have to be cut out and replaced. This is where I usually start on a repointing job—determining the number of replacement brick needed and locating brick that matches as close as possible before actually doing any cutting out.

I try to match replacement brick as close as possible for color, texture, and size so that they blend into the original work without being noticeable. If you pick used brick, be careful that they are not soft or have internal cracks. A good method for checking used or old brick is to hold one in your hand and tap lightly in the center with the head of a brick hammer. If it is soft or cracked,

it will emit a hollow or dead sound. If it is sound, it will have more of a metallic ring.

Be careful of using old salvaged brick, many of which were never intended for exterior brick walls. Some of the older brick were not burned or fired as long or with the prolonged, intense heat that face brick were. Thus they are a little softer brick and were sold for use in interior load-bearing walls of older homes. You can easily spot them as they will be very lightweight and have a soft pinkish color. They also are very porous and soak up a lot of moisture. If used in an exterior wall, they can flake off and disintegrate after a rather short period of time. I have examined houses built of soft salvaged brick that were so bad that the walls had to be completely torn down and rebuilt with new brick. Make sure that you inspect any old replacement brick to be sure that they are reasonably hard and free from cracks before using them in a wall.

A good source of used brick is the classified section in your newspaper. If you cannot locate good used brick, most brick suppliers stock new reproductions of old brick that will match pretty closely. Take a sample of the old brick that you are trying to match to the brick supplier and compare them side by side.

Mortar for Repointing

Mortar for repointing should be made of portland cement, hydrated lime, and sand. Mortar that is rich and has a too high percentage of portland cement gets very hard and will not establish a good bond to old soft brick. Mortar for repointing should never be harder than the brick it is being pointed around. What you are trying to achieve is superior bonding of the mortar to the brick, not high compressive strength. After all, the original mortar of the structure supports the load of the masonry work, not the ¾ in. to 1 in. of mortar that you repoint into the joints.

Recommended repointing mix My recommendation for a good average repointing mortar is 1 part portland cement to 2 parts hydrated lime to 8 parts washed masonry sand with enough water to make it a little stiffer than regular mortar. This is a high-lime mortar and will bond excellently to old brick and mortar joints. This mix when cured will test out to approximately 350 psi, which matches up reasonably close to the compressive strength of most older brick. If the brick and mortar joint strengths are about the same, there is a lot less chance of shrinkage or cracking from expansion in freezing and thawing weather.

The reason that a high lime mortar is preferred for repointing work is that the lime imparts the ability to reknit or reseal itself if hairline cracks occur. This unique action is known as *autogenous healing*. The hydrated lime dissolves when it comes in contact with natural rainwater and is recarbonated by carbon dioxide, which in turn reseals any hairline cracks in the mortar joints. A high-lime mortar bonds to older brick much better than modern stronger mortars.

Matching the color of the repointing mortar You should always try to match the color of the repointing mortar to the original so it does not stand out. Most building suppliers who stock masonry cements also carry dry powder tinting colors that can be added to the mix to achieve different shades of color. You'll have to experiment to arrive at the desired color

The ingredients for repointing mortar are from left to right: portland cement, hydrated lime, and sand.

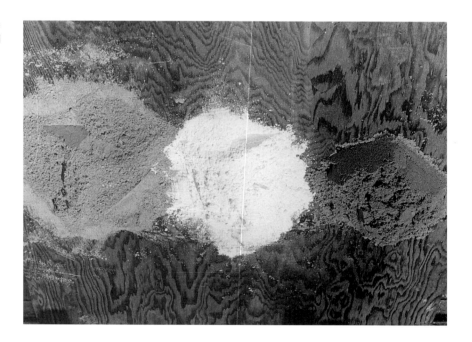

Using buff Flamingo custom color masonry cement in the repointing mix produces good results.

match. I start by mixing a small amount of mortar, adding some coloring according to the directions on the package, and testing it on a small section of the wall. I keep doing this until it looks about right. Keep in mind that mortar always looks a little darker when it is wet, so expect a little lighter color after it has dried.

There are also custom masonry cements available that are tinted in different colors. You can add a small percentage of these to the mix to create the desired color without changing the compressive strength of the mortar drastically. For example, most of the time you will find that you need a repointing mortar that has a buff or slightly yellowish color to match old faded mortar joints. An excellent masonry cement brand that I have used and recommend is called Flamingo and is made by the Riverton Corporation in Riverton, Virginia. I have used the following formula with excellent results: 1 part portland cement to 1 part Flamingo buff custom color masonry cement to 2 parts hydrated lime to 9 parts sand. You will have to specify the color of masonry cement from a color chart, as they go by code numbers on the bag. If Flamingo masonry cement is not available in your area, I am sure that you can find a substitute of equal quality.

Brickwork repointed with a buff-colored mortar blends well with the old brick and mortar.

Prehydrated mortar should stay together in a ball or lump when squeezed in the hand.

Old mortar joints should be cut out to a depth of approximately 1 in.

Prehydrating the mortar mix The purpose of prehydrating repointing mortar is to decrease the amount of shrinkage that occurs away from the edges of porous older brick in the mortar joints. To prehydrate mortar, first mix all of the ingredients with just enough water to blend them together. The mortar should be in a damp unworkable mix that will retain its form when pressed in a ball in the palm of your hand.

Let the mortar set for approximately 30 to 45 minutes, and then add only enough water to make it workable. The mortar that results is a little stiffer and drier than regular mortar and will shrink very little. For best results, don't mix big batches of repointing mortar—only enough for about two hours of work.

Cutting Out Old Mortar Joints

I recommend cutting the old mortar joints out to a depth of approximately 1 in. Use a plugging or joint chisel to do this. Plugging chisels are made in a variety of blade thicknesses to fit thin or thick mortar joints. The special shape of the angled tapered blade cleans the mortar out of the brick joints without binding in the joints or chipping the faces of the brick. The joints should be cut out as square as possible rather than in a V-shape, so that the new mortar will fill more of the area.

After the joints have been cut out, rake out loose particles with a flat, thin steel slicker. Then brush out the joints clean.

Plugging Chisel

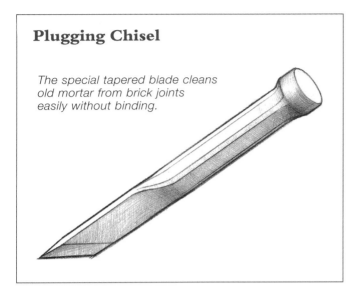

The special tapered blade cleans old mortar from brick joints easily without binding.

Before repointing fresh mortar in the joints, dampen the area of the wall to be repointed with water so the mortar will bond better and not dry too fast. I use a fine spray from a garden hose or a bucket of water and an old brush. Don't soak the wall; just make it damp.

Repointing the Joints

Regardless of the joint finish desired, I have found that a flat slicker works best for tucking mortar into the joints. I recommend always filling in the head joints first and then the bed joints to maintain the straight continuous appearance of the bed joints. Most repointed joints are pointed flush, then smoothed out with the flat blade of the slicker. However, you can tool the mortar joint with any finish you like. Be sure to do it before the mortar dries too much or the tool may leave black marks.

If you have to repoint deep mortar joints, I recommend filling in about half of the mortar joint depth, waiting until this is thumbprint hard, and then repointing the

Clean the mortar joints with a brush prior to repointing.

Fill in the head joints with the slicker tool.

Fill in the bed joints with the slicker, keeping the joints long, straight, and smooth.

A correctly repointed brick wall shows the mortar joints filled out to the edges of the brick.

rest. This cuts down on the joints shrinking, cracking, or sagging. If the wall dries out too fast, flick a little water on it with a brush from the bucket to keep it damp. It's very important to keep the old joints moist at all times if the mortar is to cure properly!

After the mortar joints have dried enough so that they will not smear, brush the wall with a soft brush to complete the job. The joints should fill out evenly to the edges of the brick with no overrun. If done properly, it usually is not necessary to clean the wall after the mortar cures.

REPAIRING CRACKS IN A MASONRY WALL

Before trying to repair a crack in a masonry wall, it is helpful to determine what caused it. Cracking occurs in different ways or patterns. By examining these patterns, you can often figure out the cause. One thing is for sure—any crack in a masonry wall is caused by movement. Basically, there are two main causes of cracks—settlement and expansion.

Settlement Cracks

Usually, settlement cracks show as relatively straight vertical lines or shears. They occur frequently when part of the foundation wall or footing settles unevenly in relation to another part of the structure. Settlement cracks are usually larger at the top and get smaller at the bottom, changing to a hairline crack at the very bottom. The tapering of the crack is the tip-off that it is a settlement crack rather than an expansion crack. If this happens on a relatively new house, I always recommend letting it sit for at least one year to allow initial settlement to complete before repairing it.

If the crack continues to move during that time, it may become necessary to saw a straight vertical slot and install an expansion joint to allow movement. Assuming that the crack is stable, cut out along the crack line to a depth of about ¾ in. with a chisel. Clean the area out well with a damp brush, and repoint with fresh mortar, leaving about a ¼-in. depression in the joint. Complete the job by caulking the surface of the crack.

A settlement crack in a brick wall. Notice how it is in a straight vertical line with the crack wider at the top and smaller at the bottom.

A typical caulked expansion joint in a brick wall.

There is a variety of very good silicone elastic caulking compounds that can be used to caulk the surface of the crack. Colored masonry caulking compounds are available at most building suppliers, which will match almost any brick, concrete block, or concrete.

Expansion Cracks

Expansion cracks usually appear in a stair-step pattern but can be vertical or horizontal. The most frequent locations of this type of cracks are over windows and doors, in gables, and where brickwork meets concrete slabs. They are usually caused by expansion and contraction due to warm and freezing weather changes. In many cases, a crack that opens up at one time of the

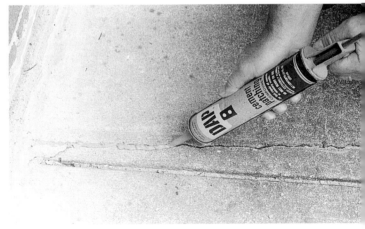

Caulk a crack in concrete with a caulk that is the same color as the concrete.

A stair-step expansion crack in a brick gable wall.

year will close up at the opposite time of the year. The crack is still there, but it is so small that it is hard to see. Steel lintels over windows and doors (especially large garage doors) will expand and cause a crack in the wall. Repair expansion cracks in the same way as settlement cracks. However, there is no guarantee that they won't appear again. Applying caulking to the surface of a crack seems to work better than anything else because it will flex and give some, especially in a crack that keeps coming back.

REMOVING OLD PAINT FROM BRICK WALLS

When restoring an older brick home to its original condition, one of the major problems is removing old paint from the surface of the brickwork. Sandblasting is very effective. The danger of sandblasting, however, particularly on old historical buildings, is that the abrasive materials used to remove the paint may actually cut away part of the surface of the brick, ruining them and allowing moisture and water to penetrate. The impact of the grit and high water pressure tends to erode the bond between the old mortar and the brick, leaving cracks where water can enter. Older brick buildings build up a protective coating over the years called a patina, which once removed makes them highly vulnerable to mois-

ture or environmental acids in the air. In my opinion, sandblasting should only be used as a last resort on older brick buildings and then by an experienced, reliable commercial sandblasting contractor.

The gentlest method I have used involves spraying with low water pressure from a garden hose and then scrubbing with detergent and a bristle brush (not a wire brush). If this does not do the job, there is an effective homemade cleaner that I have used to remove old paint without damaging the brickwork. It involves using caustic soda (lye), so be sure to wear long rubber gloves, safety glasses, and old clothing. If you get any on your skin, wash it off immediately with running water. To make the cleaner, follow these steps:

Step 1. Obtain two 8-quart plastic or rubber pails, and pour 1 gallon of water in each.

Step 2. Wearing a pair of long rubber gloves, pour 1 quart of caustic soda in one of the buckets and stir.

Step 3. Add 8 ounces of regular cornstarch to the other bucket of water, and stir the cornstarch and water together.

Step 4. Mix the water containing the cornstarch and the water with the caustic soda together, stirring constantly.

Step 5. Be sure to cover shrubbery, plants, or other objects that may be damaged with plastic before starting.

Step 6. Apply the mixture on the painted brickwork from the bottom to the top with a plastic fiber brush, being careful not get it on any exposed skin.

Step 7. It will stick like gelatin and "cook," literally eating the paint off the surface of the wall. Leave it on the wall for about one hour. If you run into a stubborn place, repeat the procedure.

Step 8. Flush the wall with water from a hose until it runs clear.

Caustic soda is available from most chemical supply houses. Regular cornstarch can be bought at the grocery store. Be sure to check in your area if you have to make provisions to trap the runoff from the wall, as some municipalities do not allow this chemical to flow into the city storm drainage system. Rinse off the area when finished to dilute any residue left on the ground or paving.

REMOVING STAINS FROM MASONRY WORK

There are a variety of stains that show up on masonry walls. If you know or can identify the type of stain, there is an assortment of cleaners on the market that are very successful in removing them. Following are some of the more frequent stains I have encountered and the chemicals that work well to remove them.

Efflorescence

Efflorescence is the most common stain that occurs on masonry walls. It is a whitish stain or scum that appears commonly on brickwork but can also be found on concrete block work. It is caused by natural salts in the mortar, brick, and concrete block. The presence of moisture causes the salt to migrate or leach to the surface as the brick or block dry out. The best prevention is to keep brick and block dry before laying them in a wall and

An efflorescence stain is caused by natural salts in masonry work that migrate to the surface of the wall.

cover the wall at the end of each workday as it is built. It also helps to keep masonry units stored off the ground on pallets and covered with tarps until ready to use.

There are two effective methods that I have used to remove efflorescence from a masonry wall. One that has been around for a long time is washing the wall down with a muriatic acid solution available from any masonry supplier. I start by soaking the wall with water from a garden hose with a spray nozzle. Then, mix a solution of 1 part muriatic acid to 10 parts water in a plastic bucket. Caution: Pour the acid into the water, not the water into the acid to prevent splashing raw acid on yourself. Always wear eye protection and long rubber gloves to protect your skin.

A grouping of products that are very effective at cleaning stains from brickwork.

Scrub the wetted wall with a stiff fiber brush dipped into the acid solution. The muriatic acid will dissolve the stain from the wall. Stubborn patches may require going over again until clean. Immediately after scrubbing, rinse the wall with water until it runs clear. This same muriatic acid solution has also been a long-time favorite for washing down new brickwork to remove mortar stains.

Smoke Stains

The Brick Institute of America has a recommendation for removing smoke stains, which is as follows: Scrub the blackened area with a scouring powder (one with bleach) and a stiff bristle brush. Rinse or wipe off with water when finished to remove the powder residue. For more stubborn stains, they suggest mixing 1 cup powered pumice (available at hardware stores or building suppliers) to 1 cup household ammonia to ½ cup water. (Warning: Never mix ammonia with any compound that contains bleach.) Stir to a thin creamy paste. Apply the mixture to the blackened area with an old paintbrush. Allow it to dry, then scrub it off with a wet scrub brush. Repeat the procedure a couple of times if necessary to remove the stains. In place of pumice, inert materials such as talc, whiting, fuller's earth, bentonite, or other clays may be used.

Other Stains and Stain Removers

A cleaning product that is very popular with professional masons for washing down and cleaning brickwork (which I recommend highly) is Sure Klean 600 Detergent™. It is manufactured by ProSoCo., Inc., and available from most masonry suppliers. It is recommended by many brick manufacturers because it is less harsh than muriatic acid to the colors in brick.

Vana Trol™ is a Sure Klean product that is excellent for cleaning new brickwork that is subject to vanadium, manganese, molybdenum, and other metal stains. Restoration Cleaner™ is another Sure Klean product that is a general-purpose acidic cleaner and is highly recommended for cleaning brick, removing atmospheric stains, and generally brightening up brickwork that is dull from exposure to the elements. It is used a lot by restoration contractors.

Sure Klean Asphalt and Tar Remover™ is an excellent product for removing asphalt and tar on brickwork. These types of stains are stubborn to remove with other types of cleaners and occur a lot when tar splashes on brickwork when built-up roofs are being installed.

Because there are so many different types of stains that can appear on masonry work, it is not possible to discuss all of them here. If you have a stain problem that defies removal, an excellent source of information is the Brick Institute of America, 11490 Commerce Park Drive, Reston, Virginia 20191. Contact the organization by mail or by calling (703) 620-0010, and the engineering staff will try to answer your questions. Before using any masonry cleaning product, be sure to read the manufacturer's recommendations on the label.

WATERPROOFING THE EXTERIOR OF A MASONRY WALL

Some types of exterior brick, block, or stone walls that are relatively soft will absorb a lot of water from rain. This water can completely penetrate the wall, causing a very damp condition inside. This is especially true for older historical brick structures. When the original appearance of the masonry work is to be preserved, this water absorption can be prevented by applying a liquid silicone product to the exterior of the wall.

Masonry silicones are sold as clear liquids and are available at building suppliers. There are two major types—water-based and solvent-based—either one of which can be applied by brushing or spraying. Clear silicones generally will not cause any change in the color of the masonry work. However, it is wise when applying them to a brick wall that has colored mortar joints to try a small area first to make sure that no bleaching takes place. The silicone is absorbed in the brick to a depth of about ⅛ in. to ¼ in. It really does not seal the wall, but it slows down water absorption by changing the contact angles between the water and the walls of the capillary pores in the masonry work. This in essence creates a negative capillary action that repels water, rather than absorbing it, and lets the wall breathe.

There are different formulas and brands of silicones available. I recommend using a 5 percent silicone resin solution, which is considered to be average. The brand is not important as long as the chemical content is the same. Any crack repair or repointing should be done at least a week before applying silicone. You can expect a silicone treatment to last from five to eight years before another application is necessary. Manufacturers recom-

A silicone repellent applied to exterior masonry surfaces can slow down water absorption.

mend spraying it on rather than applying it with a brush for best results. Be sure to follow good safety practices by wearing safety eye protection and rubber gloves when applying silicones.

TROUBLESHOOTING DAMP OR LEAKY BASEMENTS

A damp or leaky basement can be very frustrating and expensive to correct. I know from experience because I have had a problem with my own basement over the years. It can be caused by water leaking through the walls, around the area where the concrete floor meets the inside of the wall, from natural springs of water under the floor, or from condensation building up on the inside of the wall. If condensation or sweating is the problem, installing a dehumidifier is a quick cure. One of the problems with using a dehumidifier, however, is that a temperature of approximately 65°F has to be

Homemade Cement Waterproofing Paint

I have developed an inexpensive homemade formula that I have used successfully on concrete block walls for many years. All of the materials are readily available from most masonry suppliers. If you have a severe water problem that won't go away, it is best to call in a professional contractor or building inspector for advice.

To make my waterproofing paint, follow these steps:

Step 1. Mix 3 parts white portland cement to 5 parts hydrated lime to 1 part calcium chloride, adding water until it has a consistency like thick paint, in a plastic 5-gallon bucket, such as the kind that drywall compound comes in.

Step 2. Let the mixture sit for about 20 minutes, and it will thicken to a creamy consistency. Restir it, and it is ready to use.

Step 3. Dampen the surface of the wall with a fine spray from a garden hose, and brush on the paint with a whitewash brush. Alternate your brushing directions to fill any holes or voids.

Step 4. The cement paint will take about eight hours to dry.

---

maintained or the coils freeze up and do not operate. It's a good idea to install a sump pump to carry excessive amounts of water to a drain outside.

If the problem is water pressure building up against the exterior of the wall, you should try to determine where it is coming from before attempting to fix it. I would start by conducting a visual inspection around the exterior of the foundations walls to see if you can spot any obvious problem areas. Check to see if any downspouts are detached, allowing water to pool next to the foundation wall. If you do not already have them, place splash blocks on an incline under the downspouts to direct the water away from the house. You can also dig a ditch and install a drainpipe under the downspout on an incline away from the house. Sometimes the fill earth around the house sinks and allows water to collect next to the foundation wall. If this is the case, you have to raise the grade a little.

If you have inspected and corrected the possibilities mentioned above and the basement still leaks, then it is time to look below ground. If the exterior wall was not waterproofed correctly when it was built and drainpipes were incorrectly installed around the outside of the foundation wall, that could be the source of the water

problem. Almost all building codes now require flexible drainpipes around the base of foundations. Unfortunately, the only way I know to correct this is to excavate around the foundation down to the footing, rewaterproof the wall, and install flexible plastic drainpipes according to local building codes. This is the most expensive cure and seldom has to be done.

Waterproofing the interior of the wall will not be effective unless the problem is minor. Applying a good quality cement paint is a possibility. They are manufactured in a variety of colors. One product that I have used for minor dampness problems is called Thorseal. It is made by the Thoro System Products Company and available from many building suppliers around the country. Specific directions are printed on the label. I recommend cutting out and repointing any cracks in the wall with mortar before applying it. Allow a couple of days for the mortar to cure.

REPAIRING OR REPLACING A MASONRY RETAINING WALL

One of the most frequent problems in masonry walls is the failure of retaining walls. The earth and the presence of moisture exert enormous pressure on the back of the wall. Walls have to be especially designed and reinforced

An example of a masonry retaining wall that is pushed out.

to withstand this pressure. Once a retaining wall has started to push out, there is no way that you can push it back and expect it to stay in place. This is because the bond is broken between the mortar and the masonry units. The only answer is to tear it down and rebuild it, making it strong enough so that it can't push out again.

Many of the same problems that cause leaky basements also play a big part in the failure of retaining walls. I have rebuilt a number of masonry retaining walls over the years and discovered when I tore some of the old ones down that they did not have footings under them but were built on bare ground. If you have to replace a retaining wall that has pushed out or collapsed, I recommend pouring a concrete footing at least 24 in. wide, regardless of the size wall being built on it. The pressure of the earth behind the wall pressing down on the back of the footing will help restrain the walls. The footing also has to be below the frost line, the same as for any other masonry wall.

Any masonry retaining wall should be reinforced with masonry wall reinforcement, steel rods, and concrete poured in the center of block for extra support, regardless of its height. I also recommend using a higher-strength mortar than for regular brickwork, such as Type M or Type N. Always incorporate concrete or tile drains through the wall at periodic intervals to relieve any moisture backup. The area at the back of the wall, next to the footing, should have flexible drainpipes laid on a bed of gravel to allow water to drain away. I also recommend parging the back of the wall with cement mortar to waterproof it.

In addition to the standard masonry retaining wall shown in Chapter 3, I have experimented with several other designs for retaining walls where exceptional stress has been present. They provide a lot more strength than a regular wall. The first method is to build the wall, reinforcing it as discussed in Chapter 3 but adding several wing walls that are spaced 6 ft. apart on a 90-degree angle and are racked back into the bank of earth. It is not necessary to pour a footing under these as they only serve as braces, or deadmen.

The second method utilizes building concrete lintels on a 90-degree angle from the retaining wall back into the earth bank to act as deadmen. They act as anchors or restraining braces into the backfill and help prevent the

MASONRY REPAIR AND RESTORATION   **165**

A Concrete Block Retaining Wall

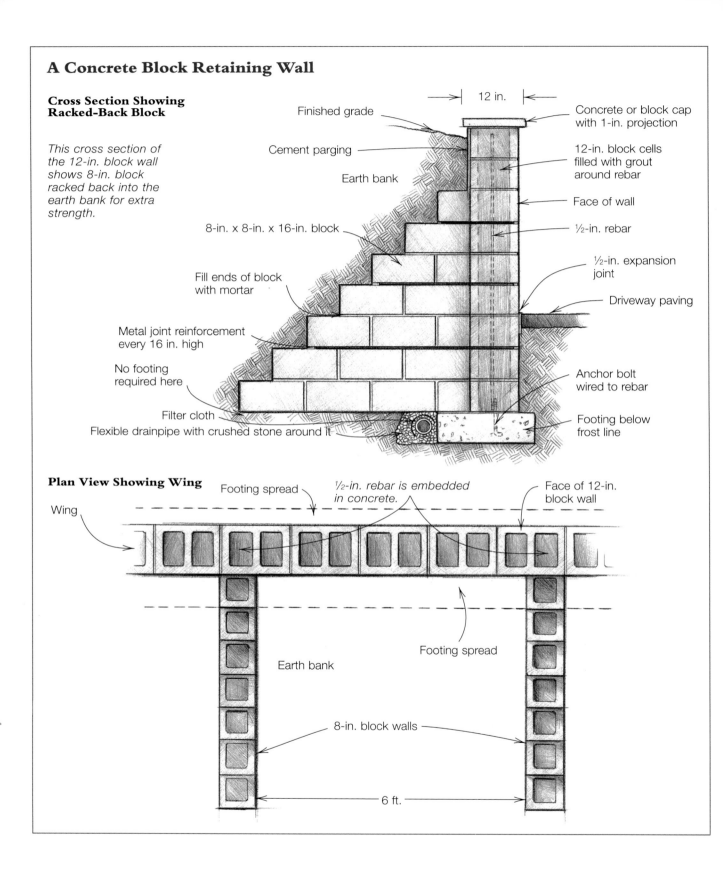

Cross Section Showing Racked-Back Block

This cross section of the 12-in. block wall shows 8-in. block racked back into the earth bank for extra strength.

12 in.

Finished grade

Cement parging

Earth bank

8-in. x 8-in. x 16-in. block

Fill ends of block with mortar

Metal joint reinforcement every 16 in. high

No footing required here

Filter cloth

Flexible drainpipe with crushed stone around it

Concrete or block cap with 1-in. projection

12-in. block cells filled with grout around rebar

Face of wall

½-in. rebar

½-in. expansion joint

Driveway paving

Anchor bolt wired to rebar

Footing below frost line

Plan View Showing Wing

Wing

Footing spread

½-in. rebar is embedded in concrete.

Face of 12-in. block wall

Footing spread

Earth bank

8-in. block walls

6 ft.

Using Concrete Block Lintels for Extra Support

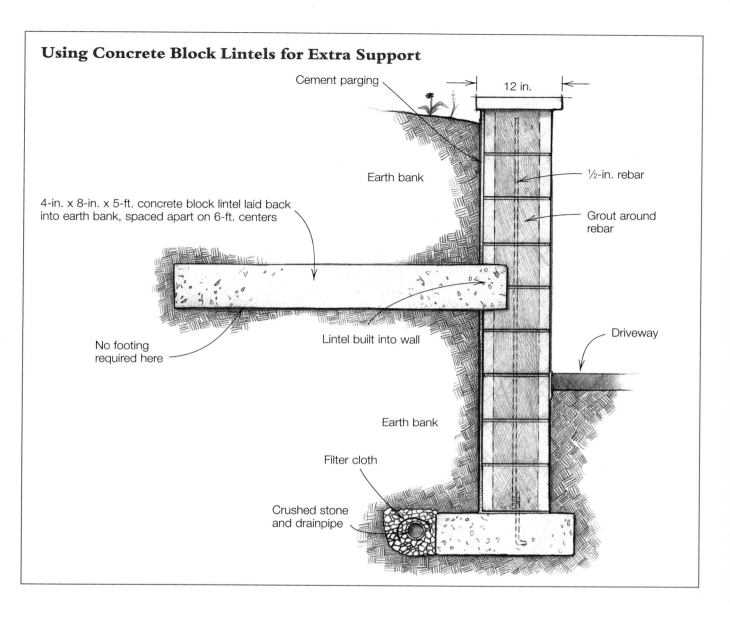

Cement parging

12 in.

Earth bank

½-in. rebar

4-in. x 8-in. x 5-ft. concrete block lintel laid back into earth bank, spaced apart on 6-ft. centers

Grout around rebar

Lintel built into wall

Driveway

No footing required here

Earth bank

Filter cloth

Crushed stone and drainpipe

wall from moving forward. I recommend spacing these on 6-ft. centers for best results. This is a more economical alternative to building block wing walls back into the fill.

REPAIRING CHIMNEYS

One of the major complaints about fireplaces that I am asked to solve is why a fireplace smokes into the house. A variety of conditions can cause this to occur, but the only way to pinpoint the problem is to conduct an on-site inspection of the fireplace and chimney. The fol-

lowing are some of the most frequent problems that I have run into that cause fireplaces to smoke and not draw properly.

Blocked Flue

The most logical thing to check first is if the flue opening in the chimney is clear all the way up. I start by opening the damper and looking up the flue. Unless the chimney is curved a lot, you should be able to see clear through to the sky above. Secondly, I go up on the roof to the top of the chimney and shine a mirror down the flue to see if it is clear. You will find that a reflecting mirror is much stronger than a flashlight, provided the sun is shining brightly.

If the chimney has not been used for a while, you may find that a bird has built a nest in the flue or that the chimney is sooty. On old chimneys where the mortar has gotten soft, occasionally part of the top of the chimney on the inside has fallen in, blocking the flue. If a bird is the culprit, I recommend lowering a burlap bag filled with some brick bats and straw down the chimney on a rope and pulling it back up to remove the obstruction. If soot or creosote has built up, try attaching a short-length chain such as a tire chain to a long medium-weight chain and lowering the assemblage down the chimney. Using a spinning motion, flail it around vigorously against the chimney walls. The chain will help to dislodge the deposits off the sides of the flue. This device is used a lot by fire companies in my area when there is a chimney fire.

If part of the top of the chimney has fallen in, the only way to repair it is to cut into the chimney at that point to remove the obstruction and rebuild the chimney. Note: If you are going to work on or clean the chimney, it's a good idea to close off the fireplace opening in the house to prevent creosote or soot from getting into the room.

Chimney Not High Enough
Another potential cause of a smoking fireplace can be that the chimney is not high enough above the roof. The recommended height of the top of the chimney should be a minimum of at least 3 ft. above the peak of the roof. The exception to the rule is if a chimney is more than 10 ft. away from the ridge of a roof, it should be at least 2 ft. above where the chimney intersects the roof line.

Tall trees that have grown higher than the house roof near the chimney can also cause downdrafts. If there are high hills or unusually deep dips in the terrain close to the house, it can cause the chimney to have downdrafts. The solution to all of these problems is to build the chimney higher. The best clue to a chimney not being

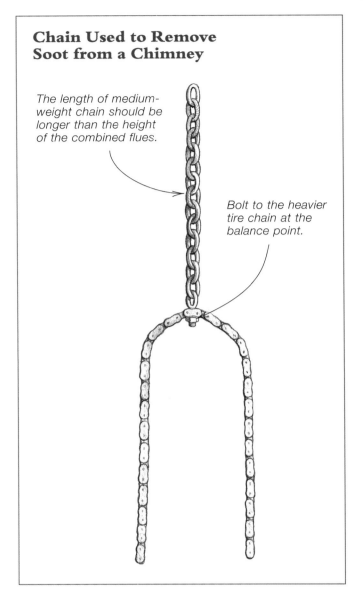

Chain Used to Remove Soot from a Chimney

The length of medium-weight chain should be longer than the height of the combined flues.

Bolt to the heavier tire chain at the balance point.

high enough is that it will smoke on a windy day but it will draw fine on an average day. I recommend adding at least an extra 4 ft. onto a chimney that does not draw properly. This usually corrects the problem.

Inadequate Draft
There are two drafts or currents of air that must be present for a fireplace to burn efficiently. One is the air that is drawn down from the top of the chimney through

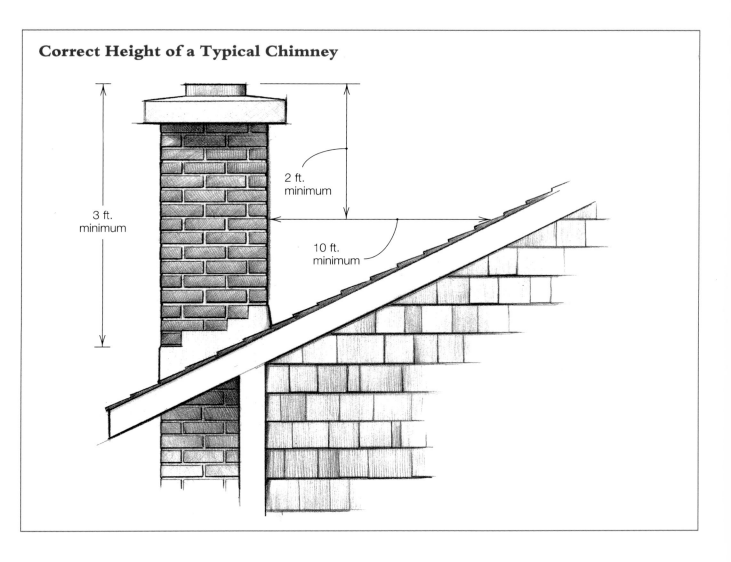

Correct Height of a Typical Chimney

3 ft. minimum

2 ft. minimum

10 ft. minimum

the flue, striking the smoke shelf in the chimney and bouncing back up the chimney. The other is the air that is drawn or sucked into the firebox from the room or from a vent in the hearth. Working together, they form a draft that carries the gases and smoke up the chimney.

If the room is too tightly sealed, opening a window a crack may provide the necessary air for a fireplace that smokes. This cure is wasteful because it also draws a considerable amount of heated air out of the house. The better solution is to install a vent to the outside, either in the fireplace hearth or as near the hearth as possible to bring in an air supply (see the left illustration on p. 170).

Shared Flue

Don't try to hook up a stove or two fireplaces in the same flue—it simply will not work. They take two separate air currents and they will conflict and not draw properly in the same flue. This situation happens in remodeling jobs sometimes, where someone decides to open up an old fireplace on another floor and hook it up to the same flue as the main fireplace.

Deteriorating Mortar Wash Cap

The top of many chimneys is sealed off with a cement mortar wash on an angle of approximately 45 degrees. Over a long period of time, these mortar caps can dete-

How Draft Works in a Chimney

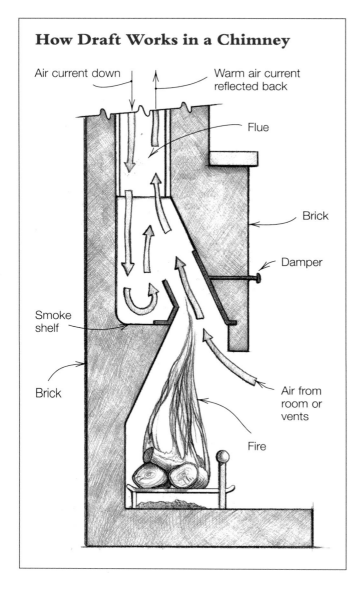

Air current down

Warm air current reflected back

Flue

Brick

Damper

Smoke shelf

Brick

Air from room or vents

Fire

Cement Mortar Wash on Chimney

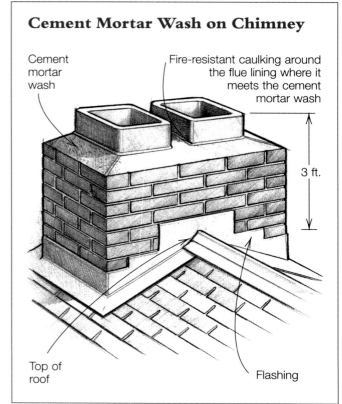

Cement mortar wash

Fire-resistant caulking around the flue lining where it meets the cement mortar wash

3 ft.

Top of roof

Flashing

Chimney tops where the mortar joints have eroded can be cut out and repointed following the same procedures as repointing normal masonry work. However, if the top of the chimney is cracked all the way down to the roof and the brick have deteriorated, there is no use kidding yourself. You may as well take it down to where it is solid, and rebuild it, including new flashing.

As I stated at the beginning of this book, there is not only a lot of satisfaction to be gained in being able to do your own masonry work but also in being able to increase the value of your property. There is a real sense of pride and accomplishment in taking an old masonry building and restoring it to its original condition. I have enjoyed sharing my knowledge of masonry work with you throughout this book and wish you the best of success in your masonry endeavors.

riorate, crack, or crumble, allowing water to penetrate the top of the chimney. To repair, remove the old mortar carefully with a flat blade chisel and replace with fresh cement mortar such as Type M or S, angled on a slope with the trowel blade. I have found that it helps to cover the fresh mortar with a wet piece of burlap to prevent it from drying out too fast. After the mortar has cured, run a caulking bead of a heat-resistant caulk around the point where the flue lining meets the mortar wash to allow for expansion.

This chimney has been torn down to the roof line.

The same chimney completely rebuilt.

The back view of an older brick house before restoration.

The front view of the same house after being restored.

GLOSSARY

Admixture A substance or chemical other than portland cement, aggregates, and water added to mortars or concrete to achieve a special condition such as workability, air entrainment, water retention, and acceleration or retarding of setting time.

Adobe brick Brick that is made from clay and placed in a mold to dry and cure in the sun. In ancient times, straw was added to the clay to reinforce it.

Aggregate A coarse material, such as sand, stone, gravel, or manufactured substitutes, that is used in mortar and concrete to bind the mix together and increase its strength.

Air-entraining agent An additive that traps and stabilizes air bubbles in concrete creating small voids or cushions, which in turn permit concrete to resist freeze-thaw cycles.

ASTM (American Society for Testing and Materials) An organization that establishes standards for building materials.

Autogenous healing The ability of mortar joints with a high lime content to reknit or reseal hairline cracks when the lime comes in contact with moisture.

Backfilling A process of filling in earth around a foundation after it has been built up to a finished grade level.

Backing up A process of laying the inside masonry units of a composite wall after the face side has been built. An example is a brick wall backed up with concrete block.

Basket-weave bond A bond used for paving work where two full brick are laid in one direction and two are laid in the opposite direction.

Batter The recession of successive courses of masonry on a gradual incline; the opposite of a corbel.

Beam pocket A recess left in a masonry wall, usually at the top of a wall or on a pier, on which the supporting beam that carries the floor joists rests.

Bearing wall A masonry wall that supports an imposed weight other than its own, such as a wall that supports floor joists.

Bed joint The horizontal bed of mortar that a brick is laid on.

Bond This term has three different meanings as applied to masonry work: (1) tying together various parts of a masonry wall by lapping units laid in mortar one over another or by connecting with metal ties; (2) the varying pattern formed by the exposed faces of masonry units; (3) the adhesion between mortar, grout, or concrete and masonry units or reinforcement steel.

Brick Structural units of kiln-burned clay and shale that are generally rectangular in shape.

Brick, engineered Brick that measure $3\frac{5}{8}$ in. wide by $2\frac{3}{4}$ in. high by $7\frac{5}{8}$ in. long. The engineered brick is more commonly called an oversize brick in the masonry trade.

Brick, Norman Brick that measure $3\frac{5}{8}$ in. wide by $2\frac{1}{4}$ ($2\frac{2}{3}$) in. high by $11\frac{5}{8}$ in. long.

Brick, Roman Brick that measure $3\frac{5}{8}$ in. wide by $1\frac{5}{8}$ in. high by $11\frac{5}{8}$ in. long.

Brick, standard Brick that measure $3\frac{5}{8}$ in. wide by $2\frac{1}{4}$ ($2\frac{2}{3}$) in. high by $7\frac{5}{8}$ in. long.

Brick, SW A term for brick that are burned very hard and recommended for paving work. They are classified as "severe weathering" brick.

Brick, utility Brick that measure 3⅝ in. wide by 3⅝ in. high by 11⅝ in. long. Also commonly called economy brick in the masonry trade.

Brick bat A trade term for half a brick.

Brick cavity wall A masonry wall that has an air space between the front and the backing.

Brick kiln A furnace-type oven or heated enclosure used for firing or burning brick to make them hard.

Brick set chisel (also known as a blocking chisel) A chisel with a 4-in.-wide blade used in masonry work to cut more exact pieces of brick or block.

Brick tongs A metal carrier that is held in the hands and is used to pick up and transport from eight to ten brick to the work area.

Brick veneer wall The outside brick facing that covers or encloses a wall built of other materials such as 4-in.-wide brick tied to a 2x4 wood frame.

Burning the joints Tooling mortar joints when they have dried too much, which causes black marks to appear from the metal striking tools.

Calcium chloride A chemical additive used to accelerate the setting time of mortar or concrete.

Chase A trade term for a vertical or horizontal recess or slot built into a masonry wall, usually to receive plumbing pipes or electrical conduits.

Chimney pad A concrete footing for a chimney that is heavier than a normal footing.

Chimney wash The angled slope of mortar on top of a chimney that seals it, helps any water to drain off, and promotes a smoother flow of air over the top.

Clinker A burned material that is fused into a mass when it cools. A clinker brick is one that was burned too long and under too intense heat, resulting in a distorted size and black fused materials on the face.

Closure brick The last brick or block laid in the wall, usually in the center of the course.

CMU The abbreviation for concrete masonry units or block.

Collar joint The mortar joint between two tiers of masonry units, such as between a brick wall backed up with concrete block.

Colonial lip A lipped edge at the top face side of colonial brick.

Colonial pink A light pink-colored brick similar to brick made in the colonial period.

Common bond (also known as American bond) An arrangement of brick in the stretcher position with a header course on the fifth, sixth, or seventh course to bond the wall together.

Composite masonry wall Any masonry wall consisting of two separate tiers of masonry units bonded together with brick headers or metal wall ties, such as brick backed up with concrete block.

Concave joint A half-round impression formed in mortar joints by using a convex jointer.

Concrete block A hollow or solid block made from portland cement and aggregates for the purpose of building masonry walls.

Concrete prescription mix A method of placing an order for concrete based on the compressive pounds per square inch (psi) strength designation after it has cured for a 28-day cycle; for example, a 3,500 psi footing mix.

Corbel The projection or racking out of brick in a masonry wall to form a shelf or ledge; the opposite of a batter.

Corner poles Manufactured metal or fiberglass poles that replace the brick corner as a guide when building masonry walls to a line.

Course A trade name for a row or layer of masonry units such as brick or block laid in mortar.

Course counter spacing rule A 6-ft. folding rule used by masons to lay off individual courses of brick that are not made to the 4 in. modular-grid scale.

Course rod or story pole A wooden pole or rod upon which are marked all of the individual masonry courses to scale, such as windows, doors, and sills, for a typical story of a house or building.

Crazing Hairline cracks that appear in the surface of cured concrete and resemble cracked eggshells; usually caused by the concrete drying too fast or by finish-troweling when there is too much surface water present on the concrete.

Cream The cement paste that is worked up to the top of fresh concrete by floating to allow finish troweling.

Cross joint A trade term for a mortar head joint particularly when it is applied to the longest side of a brick.

Crowding the line (also called hard to the line) A trade term for laying brick or block against the line so that it pushes the line out of alignment.

Cube of brick Brick that are banded together with steel bands to allow moving and transporting them by truck. A standard cube of brick usually contains 500 brick with a few extra for breakage.

Cubic yard The common volume measure for concrete, which contains 27 cu. ft.

Cupping mortar A method of picking up mortar from a mortarboard by rolling and cupping it on a brick trowel.

Deadman An anchor or form of brace usually buried or built into the ground to help strengthen, support, or stabilize a masonry wall. An example would be concrete block lintels built into a block retaining wall and extending back into the earth fill behind it to prevent it from moving. It can also mean a prop or post to attach a line to when laying brick to the line.

Double jointing A process of applying mortar to both ends of a closure brick as well as to the ends of the adjoining brick already laid in the wall.

Draft In a chimney or fireplace, the current of air created by the variation in pressure resulting in differences in weight between hot gases in a flue and the cooler air outside a chimney.

Drowning the mortar A trade term meaning to add too much water to mortar, making it runny or weak.

Dry bonding A process of laying out brick without using any mortar to establish the bond. The end of the forefinger is commonly used as a spacer between each brick to allow for mortar head joints.

Dusting A loose, dusty condition that results when the surface is left to set too long before finish troweling. It can also result if the surface water on fresh concrete freezes before it cures.

Dutch corner A method of starting a brick corner by laying a 6-in. piece (also called a three-quarter) of brick on or against the corner return brick.

Ears The projecting ends of a regular stretcher concrete block where the mortar head joint is applied.

Edging The process of using a rounded metal edging tool to round off the edges of concrete, such as on a curb or sidewalk to prevent chipping of the edges.

Efflorescence A whitish deposit or stain on masonry work caused by the leaching of soluble salts from within the wall due to the presence of moisture in the wall.

English bond A brick bond or pattern in which there are alternating courses of stretchers and headers throughout the wall.

English corner A method of starting a brick corner by laying a 2-in. brick piece (also called a plug) against the corner return brick.

Face shell bedding A method of applying mortar bed joints only on the outside webs of a concrete block.

Fall A trade term for pitch or slope.

Fat mortar A mortar that has a rich content of cement, making it sticky.

Flemish bond A brick bond in which a stretcher and a header alternate on the same course. The header on every other course should be centered over the stretcher below.

Floating A process of rubbing or troweling the surface of freshly placed concrete with either a wood or metal tool to bring the cement paste to the surface for finish troweling.

Fluted block Special manufactured concrete block where the exposed faces have vertical projecting ribs with slots in between for creative or ornamental masonry appearances.

Frog A depression or indentation in the bed surface of a brick; used for design or as an aid for locking the brick more securely in the mortar bed joint.

Full header A header course of brick that consists of all brick laid in the header position.

Furrowing A process of forming indentations in the center of the mortar bed joints with the point of a brick trowel to distribute it evenly prior to laying brick on it.

Garden wall bond An adaptation of the Flemish bond where two stretchers and a header brick alternate on the same course. This is known as a double-stretcher garden wall bond. If three stretchers are used with a header, it is known as a triple-stretcher garden wall bond. This type of bond creates diamond patterns in a brick wall.

Grapevine joint A continuous line formed in the center of mortar joints by using a grapevine metal jointing tool. The line can be wavy or straight.

Green masonry or concrete work A trade term used to describe mortar joints or concrete before it has cured and become hard.

Grooving A process of using a metal tool called a groover that has a rounded projected edge on the bottom and in the center of the tool to form a straight groove in concrete sidewalks or slabs where control or expansion joints are located.

Half-lap pattern The lapping of one masonry unit halfway over the one below it in a masonry wall.

Hawk A flat piece of metal or wood with a handle attached to it for holding mortar when plastering or parging.

Header high A trade term used to describe the height of a brick wall where the brick header course is to be laid.

Head joint The mortar joint between the ends of brick in a masonry wall.

Herringbone bond A brick bond in which the arrangement of the brick are in a zigzag pattern with the end of each brick laid on right angles against the side of a second brick. Generally the herringbone pattern is laid on a 45-degree angle off of a border.

Hog in the wall A trade term describing a masonry wall that is built to the same height on both ends but one end contains one more course than the other. This is usually caused by one mason forming larger mortar bed joints under the brick than the mason on the opposite end. Using a course rod prevents this from happening.

Honeycombing A common problem in concrete caused by incomplete filling out against the edge of forms, which leaves a series of voids like the cavities in a bee's honeycomb.

Humored A trade term indicating a gentle adjustment of masonry work by a mason so that differences are not evident to the eye. Also sometimes called fudging.

Hydrated lime Quick lime to which sufficient water has been added to convert the oxides to hydroxides. This is the type of lime required for masonry mortars.

Hydration The formation of a compound by combining water with another substance to create a chemical action such as in mortar or concrete; for example, the reaction between the portland cement and water.

Initial set The initial drying stage of mortar to the brick or block in the mortar joint. This initial set should not be broken by shifting the masonry unit or the mortar joint may leak.

Jack over jack A trade term describing the laying of one brick over the one beneath without overlapping and with all of the vertical head joints in a plumb line, such as in a stack bond.

Jointer (also known as a striking iron) A metal tool used to tool and finish mortar head and bed joints to a desired finish.

Knee board Scrap piece of plywood laid on the surface of freshly floated concrete that the finisher can kneel on when troweling the concrete without sinking into the concrete.

Lean mortar A mortar that has less than the required content of cement in the mix, making it sandy or weak.

Level When an object is placed in a true horizontal plane or alignment.

Line block A wooden or plastic block that is attached to a masonry corner or lead for the purpose of holding a line in position as a guide for laying individual masonry courses.

Line pin A steel pin used to wrap a stretched mason's line around when laying brick or concrete block to the line.

Lintel A horizontal structural member or beam placed over a wall or opening to carry or support the weight of masonry work, such as over a door or window opening.

London brick trowel A popular pattern brick trowel that has a narrow blade and a diamond-shaped heel.

Masonry cement A prepared cement that contains all of the necessary cementitious ingredients, including admixtures, so that only sand and water need to be added to form mortar. Masonry cement is made in a variety of strength types to meet job requirements.

Masonry lead A part of a masonry wall that is built before the wall for the purpose of attaching a line to it to use as a guide for laying individual courses of brick or block.

Masonry wall reinforcement A rigid wire joint reinforcement in either a metal truss or ladder design that is used in mortar bed joints in masonry walls to bond them together or to provide additional strength.

Module In masonry work, the unit of measure used as a standard grid that equals 4 in.

Modular masonry Masonry construction where the overall size of the masonry walls are based upon the modular unit of 4 in. or multiples thereof.

Modular spacing rule A 6-ft. folding scale rule used by masons to lay off individual courses of masonry units such as brick or block based on the 4-in. module or grid.

Mud A trade name for mortar or plaster.

Muriatic acid A form of hydrochloric acid that is diluted with water and used for cleaning mortar and for removing certain stains from the face of masonry walls.

Nominal A dimension for a masonry unit that includes an allowance for mortar joints. An example is a concrete block that is 15⅝ in. in actual length but equals 16 in. when a standard ⅜-in. mortar head joint is added to the end of the block.

Parging Cement masonry mortar used to plaster block walls to seal and waterproof them.

Patina A sheen or buildup of a protective covering or deposit on the face of brickwork resulting from exposure to the atmosphere and weather conditions over a long period of time. It is particularly evident on historical brickwork.

Philadelphia brick trowel A brick trowel that has a wide blade and square heel design.

Pier A brick column used in foundations for supporting beams, as supports under porches, or for strengthening garden walls, entrances at driveways, and a variety of similar projects.

Pilaster A brick pier that is built as part of a wall but projects out or is offset from the face of the wall. If the pier projects on both sides of the wall, it is known as a double pilaster.

Plasticizer A water-reducing additive that allows as much as 15 percent reduction in water volume to achieve a specified slump requirement. It tends to increase the strength of concrete.

Plug A 2-in. piece of brick.

Plugging chisel (also known as a joint chisel) A tapered blade chisel used in masonry work for cutting out mortar head and bed joints.

Plumb A true vertical line or perpendicular alignment, such as a plumb line.

Plumb rule A trade name for a bricklayer's level that exceeds 42 in. in length. The traditional bricklayer's plumb rule is 48 in. long.

Pointing trowel A smaller version of a brick trowel that is used to point up holes in mortar or patch masonry work.

Portland cement The fine, grayish powder formed by burning limestone, clay, or certain shales and then grinding the resulting clinker. The result is a cement that hardens when mixed with water and which is used as a base for mortars and concretes. Portland cement is a grade not a brand of cement.

Portland cement/lime mortar Masonry mortar that is composed of portland cement, hydrated lime, and sand mixed with water.

Preblended cement A prepared mortar mix that only requires adding water to form mortar.

Proprietary compound In masonry work, a chemical compound protected by a patent, copyright, or trademark, which is used to clean masonry work.

psi An engineering term meaning compressive strength measured in pounds per square inch.

Reinforced brick wall A brick wall that has steel reinforcement rods and concrete grout between the masonry backing material for extra strength, such as in a retaining wall.

Repointing (also known as tuckpointing) The process of cutting out and repointing old mortar joints in masonry work.

Retaining wall A masonry wall that is constructed to restrain earthen fill. If built of brick or concrete block, it should be reinforced in the center with steel rods and concrete grout.

Return The turn and continuation of a masonry wall in another direction from which it had been started, such as on a 90-degree-angle corner.

Reveal The portion or return end of a masonry wall or jamb that is visible from the face of the wall back to the window or door frame.

Rowlock (also spelled rolock) A brick that is laid in position on its longest narrowest edge in the mortar bed joint; commonly used as a windowsill or for capping off a brick wall.

Rule-of-thumb estimating A trade term referring to a method of estimating masonry materials for a job based on known quantities. This is not an exact measurement but it works well for jobs up to the size of a house and allows for normal material waste factors.

Running bond (also called an all-stretcher bond) A brick bond consisting of all full brick laid lapped over each other.

Sailor A brick that is laid in the mortar bed joint in a vertical position with its largest side facing the face of the masonry wall.

Scaling Flaking off of the surface of concrete after it has hardened, usually caused by freezing and thawing or letting the concrete dry too long before finishing.

Screeding The process of using a straightedge board or metal tool called a screed to level off concrete even with the top of a form. This is done to bring the cement paste to the surface of the mix to cover all of the aggregates and allow desired finishing of the concrete.

Serpentine wall A masonry wall, such as a brick wall, built in a continuous, repeating, back-and-forth curving line, off of an established center line. Serpentine walls provide superior lateral horizontal strength due to the resistance of the curve acting against an opposite curve next to it.

Set of a trowel The angle of the rise of the trowel handle in relation to the blade when laying in flat position on the blade.

Shove joint The procedure of shoving one brick against another in the mortar bed to form a head joint on the ends.

Siltation test A method of determining the amount of loam and impurities in sand by placing some sand in a glass jar, adding water on top, and measuring the loam that settles on top of the sand overnight.

Skatewheel joint raker A tool used in masonry work to rake out mortar joints to a predetermined depth.

Slack to the line A trade term for laying brick or block more than the standard ¹⁄₁₆ in. back from the line.

Slicker joint A flat smooth finish on the surface of a mortar joint made by a flat blade metal slicker tool.

Slump test A test performed on fresh concrete to determine if the concrete has the correct amount of cement and water in the mix.

Spall A small chip of brick or block.

Splitrock (also known as splitblock) A solid concrete block product made of a coarse stone aggregate in which the split face is used as the face of the finished wall.

Spotting a brick The process of laying a brick to establish a specific point without the aid of a line, such as when laying out a wall on the first course.

Spreading mortar (also called stringing mortar) The procedure of cutting and spreading mortar on a brick wall to lay brick in.

Stack bond A bond in which masonry units are laid directly over the others below with the mortar head joints all in one plumb vertical line.

Story high A trade term meaning one-story height of a house, usually 8 ft.

Striking the joints (also known as tooling the joints) The procedure of tooling and finishing the mortar joints between the courses of brick, block, or masonry units.

Tailing a lead The procedure of aligning the ends of corner brick diagonally with a level.

Three-quarter A 6-in. piece of brick.

Thumbprint hard A trade term describing using the end of the thumb as a gauge to determine if the mortar joints have dried enough to tool.

Trig A metal clip that is attached to a line on brick or block laid in the center of a masonry wall to keep the line from sagging or being blown out of alignment.

Trig brick A brick set in mortar ahead of the line in the center of the course, which the metal trig is attached to to keep the wall at the correct height and plumb.

V-joint A V-impression formed in mortar joints using a V-shaped jointing tool.

Water/cement ratio The proportion of water to portland cement in concrete, usually stated in gallons of water per 94-pound bag. The lower the water/cement ratio, the stronger the concrete will be.

Wythe The thickness of masonry units in a vertical section of wall, such as a 4-in. brick wall backed up by an 8-in. block wall.

INDEX

Mortar mixes:
    masonry cement, 12
    portland-cement/lime, 12
Mortar mixing boxes, discussed, 35
Mortarboards, discussed,
    35-36

P

Paving bonds:
    basket weave, 49-51
    herringbone, 49-51
    running bond, 49-51
    stack bond, 49-51
Pilasters, discussed, 58
Plugs, defined, 45
Plumb rules. See Levels.
Pointing trowels, described, 24
Portland cement, as concrete component, 133-34
Portland-cement/lime mortar mix,
    described, 12

R

Repointing, mortar for:
    color of, 154-55
    mix for, 154
    prehydrating, 156
Retaining walls, repairing, 164-67
Running bonds:
    half-lap, 40-41
    one-third lap, 40, 42

S

Sand:
    cost of, 13
    estimating for concrete block, 117
    for mortar, 12-13
    for walkways, 109
Scaling, in concrete, 151
Screen block, defined, 11
Settlement cracks, repairing, 158-59
Sidewalks:
    curing concrete surface of, 150
    edging, 147
    finishing concrete surface of,
      148-50
    floating concrete surface of, 146
    grooving, 147-48
    pouring concrete for,
      143-44

preparing forms for, 142-43
screeding concrete for,
    144-46
Sill rods, laying out, 72-73
Skatewheel joint rakers, described,
    30-31
Slickers, described, 30
Spacing rules:
    brick mason's, using, 67-70
    choosing, 26-27
    modular, using, 70-71
    types of, 26
Spalling off, avoiding, 147-48
Splitblock, defined, 11-12
Splitrock. See Splitblock.
Squares, discussed, 31
Stack bonds, discussed, 48
Stain removers, for masonry work,
    162-63
Steel joint reinforcements, defined,
    15, 16
Stone screenings, for walkways, 109
Strap anchors, defined, 15
Striking irons. See Jointers.

T

Tempering, of mortar, 82-83
Tile hammers, discussed, 24
Tool bags, discussed, 34
Tree wells, wall for, 57, 60
Trig brick:
    discussed, 33
    setting, 96-97
Troweling, of concrete surfaces,
    148-50
Trowels. See Bricklayer's trowels.
    Pointing trowels.

U

Utility wire cutters, discussed, 34

V

V-jointers, described, 29

W

Walkways, brick, 106-109
    filling in with stone screenings
      and sand base of, 108, 109
    installing border forms of, 107,
      109

    laying paving brick of, 108, 109
    preparing base of, 106
    sweeping sand in joints of, 109
Wall ties:
    corrugated, 14
    steel joint reinforcements, 15, 16
    strap anchors, 15
    Z-ties, 15
Walls:
    brick positions in,
      header, 39
      rowlock, 39
      sailor, 39
      shiner, 39
      soldier, 39
      stretcher, 39
    for flower bed, 60
    for mailbox, 61
    for planter over drain, 61
    removing paint from,
      160-61
    repairing cracks in, 158-60
    for tree well, 57, 60
    types of,
      brick and block, 52
      brick cavity, 53-55
      brick veneer, 52-53
      concrete block or brick
        retaining, 55
      reinforced, 54-55
      serpentine, 55-57
      solid brick, 52
    waterproofing, 163
Waterproofing:
    basements, 163-64
    exterior masonry walls, 163
    foundations, 128-31
Waterproofing paint, for
      concrete block walls, 164
Wheelbarrows, discussed, 34

Z

Z-ties, defined, 15

PUBLISHER: Jim Childs

ACQUISITIONS EDITOR: Steve Culpepper

EDITORIAL ASSISTANT: Carol Kasper

EDITOR: Nancy N. Bailey

LAYOUT ARTIST: Suzie Yannes

PHOTOGRAPHER: Richard Kreh

ILLUSTRATOR: Robert La Pointe

INDEXER: Diane Sinitsky

TYPEFACE: Stone Serif

PAPER: 70-lb. Utopia Two Matte

PRINTER: R. R. Donnelley, Willard, Ohio